Statistical evaluation of
mutagenicity test data

Statistical evaluation of mutagenicity test data

UKEMS sub-committee on guidelines for mutagenicity testing. Report. Part III

EDITOR
David J. Kirkland

ASSOCIATE EDITORS
Garry A.T. Mahon
Colin F. Arlett
W. David Robinson
Chris Richardson
David A. Williams
Dennis Cooke
David P. Lovell
Dennis O. Chanter
Robert D. Combes
Mervyn R. Thomas

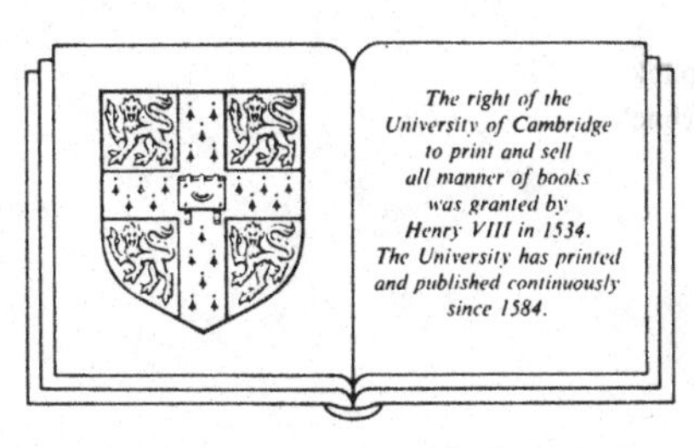

CAMBRIDGE UNIVERSITY PRESS
Cambridge
New York Port Chester
Melbourne Sydney

CAMBRIDGE UNIVERSITY PRESS
Cambridge, New York, Melbourne, Madrid, Cape Town, Singapore, São Paulo

Cambridge University Press
The Edinburgh Building, Cambridge CB2 8RU, UK

Published in the United States of America by Cambridge University Press, New York

www.cambridge.org
Information on this title: www.cambridge.org/9780521366052

First published 1989
This digitally printed version 2008

A catalogue record for this publication is available from the British Library

Library of Congress Cataloguing in Publication data
Statistical evaluation of mutagenicity test data/editor, David J. Kirkland;
associate editors, Garry A.T. Mahon . . . [et al.]; steering group, David J.
Kirkland . . . [et al.].
p. cm.
Includes bibliographies and index.
ISBN 0 521 36605 4
1. Mutagenicity testing—Statistical methods. I. Kirkland, David J. II.
Mahon, Garry A.T.
RA1224.4.M86S73 1989
616′.042—dc19 88–36461
CIP

ISBN 978-0-521-36605-2 hardback
ISBN 978-0-521-04814-9 paperback

CONTENTS

STEERING GROUP

David J. Kirkland, Microtest Research Ltd, University Road, Heslington, York YO1 5DU, UK

Bryn A. Bridges, MRC Cell Mutation Unit, Sussex University, Falmer, Brighton BN1 9RR, UK

Brian J. Dean, Glaisdale, Main Street, Low Catton, Stamford Bridge, York YO4 1EA, UK

Susan Hubbard, Health and Safety Executive, Magdalen House, Stanley Precinct, Bootle, Merseyside L20 3QZ, UK

James M. Parry, School of Biological Sciences, University College of Swansea, Singleton Park, Swansea SA2 8PP, UK

David A.H. Pratt, Cleveland, Blind Lane, Bourne End, Buckinghamshire SL8 5LF, UK

Eileen Rubery, Department of Health and Social Security, Hannibal House, Elephant and Castle, London SE1 6TE, UK

Dennis O. Chanter, Department of Statistics, Huntingdon Research Centre, Huntingdon PE18 6ES, UK

Michael J. Godley, ICI plc, Pharmaceuticals Division, Mereside, Alderley Park, Macclesfield, Cheshire SK10 4TG, UK

David P. Lovell, The British Industrial Biological Research Association, Woodmansterne Road, Carshalton, Surrey SM5 4DS. UK

CONTRIBUTORS

Rosario Albanese, ICI Pharmaceuticals Division, Mereside, Alderley Park, Macclesfield, Cheshire SK10 4TG, UK

Jeffrey A. Allen, Department of Mutagenesis and Cell Biology, Huntingdon Research Centre, Huntingdon PE18 6ES, UK

Gillian Amphlett, Glaxo Group Research Ltd, Ware, Hertfordshire SG12 0DJ, UK

Diana Anderson, The British Industrial Biological Research Association, Woodmansterne Road, Carshalton, Surrey SM5 4DS, UK

Colin F. Arlett, MRC Cell Mutation Unit, Sussex University, Falmer, Brighton, BN1 9RR, UK

John C. Asquith, Toxicol Laboratories Ltd, Bromyard Road, Ledbury, Herefordshire HR8 1LH, UK

Angus Bateman, Retd

James Bootman, Life Science Research, Elm Farm Laboratories, Occold, Suffolk, IP23 7PX, UK

Dennis O. Chanter, Department of Statistics, Huntingdon Research Centre, Huntingdon PE18 6ES, UK

M. Gillian Clare, Shell Research Ltd, Sittingbourne Research Centre, Sittingbourne, Kent ME9 8AG, UK

Geoffrey M. Clarke, School of Mathematics and Physics, Sussex University, Falmer, Brighton BN1 9RR, UK

Jane Cole, MRC Cell Mutation Unit, Sussex University, Falmer, Brighton BN1 9RR, UK

Robert D. Combes, School of Biological Sciences, Portsmouth Polytechnic, King Henry I Street, Portsmouth PO1 2DY (current address: Inveresk Research International Ltd, Musselburgh EH21 7UB, UK)

Dennis Cooke, School of Mathematics and Physical Sciences, Sussex University, Falmer, Brighton BN1 9RR, UK

Caroline J. Doré, Division of Medical Statistics, MRC Clinical Research Centre, Watford Road, Harrow, Middlesex HA1 3UJ, UK

Robert Ferguson, ICI Pharmaceuticals Division, Mereside, Alderley Park, Macclesfield, Cheshire SK10 4TG, UK

Martyn B. Ford, School of Biological Sciences, Portsmouth Polytechnic, King Henry I Street, Portsmouth PO1 2DY, UK

R. Colin Garner, Cancer Research Unit, University of York, Heslington, York YO1 5DD, UK

David Gatehouse, Glaxo Group Research Ltd, Ware, Hertfordshire SG12 0DJ, UK

Michael H.L. Green, MRC Cell Mutation Unit, University of Sussex, Falmer, Brighton BN1 9RR, UK

Michael J.R. Healy, London School of Hygiene and Tropical Medicine, Keppel Street, London WC1E 7HT, UK

Leigh Henderson, Huntingdon Research Centre Plc, Huntingdon PE18 6ES, UK (current address: Unilever Research, Colworth Laboratory, Colworth House, Sharnbrook, Bedford MK44 1LQ, UK)

James Hepworth, Computer Services, University of Southampton, Highfield, Southampton SO9 5NH, UK

David P. Lovell, The British Industrial Biological Research Association, Woodmansterne Road, Carshalton, Surrey SM5 4DS, UK

Garry A.T. Mahon, 96 rue des Maraîchers, L-2124 Luxembourg

Douglas B. McGregor, Inveresk Research International Ltd, Musselburgh, EH21 7UB, UK (current address: Boehringer Ingelheim Pharmaceuticals Inc., 90 East Ridge, PO Box 368, Ridgefield, Connecticut 06877, USA)

Brian Middleton, Statistics and Information Unit, ICI Pharmaceuticals Division, Mereside, Alderley Park, Macclesfield, Cheshire SK10 4TG, UK

Ian de G. Mitchell, Beecham Pharmaceuticals, Research Division, Honeypot Lane, Stock, Ingatestone, Essex CM4 9PE, UK

Anthony K. Palmer, Huntingdon Research Centre Plc, Huntingdon PE18 6ES, UK

David G. Papworth, MRC Radiobiology Unit, Harwell, Didcot, Oxfordshire OX11 0RD, UK

Barry Phillips, The British Industrial Biological Research Association, Woodmansterne Road, Carshalton, Surrey SM5 4DS, UK

Chris Richardson, ICI Plc, Central Toxicology Laboratory,

Alderley Park, Macclesfield, Cheshire SK10 4TJ, UK (current address: Life Science Research, Elm Farm Laboratories, Occold, Suffolk IP23 7PX, UK)

Margaret Richold, Unilever Research, Colworth Laboratory, Colworth House, Sharnbrook, Bedford MK44 1LQ, UK

W. David Robinson, Research Computing and Statistics Departments, Glaxo Group Research Ltd, Ware, Hertfordshire SG12 0DJ, UK

David W. Salt, Department of Mathematics and Statistics, Portsmouth Polytechnic, Hampshire Terrace, Portsmouth PO9 2EG, UK

John R.K. Savage, MRC Radiobiology Unit, Harwell, Didcot, Oxfordshire OX11 0RD, UK

David M. Smith, Applied Statistics Research Unit, Mathematical Institute, Cornwallis Building, University of Kent, Canterbury, Kent CT2 7NF, UK

Michael T. Stevens, Fisons Plc, Pharmaceutical Division, R and D Laboratory, Bakewell Road, Loughborough, Leicestershire LE11 3SD, UK

Mervyn R. Thomas, Dunelm House, The Street, Doddington, Kent, UK

David J. Tweats, Genetic and Reproductive Toxicology Department, Glaxo Group Research Ltd, Ware, Hertforshire SG12 0DJ, UK

David A. Williams, University of Edinburgh, Department of Statistics, James Clerk Maxwell Building, The King's Buildings, Mayfield Road, Edinburgh EH9 3JZ, UK

PREFACE

D.J. KIRKLAND

1 OBJECTIVES

In March 1982 the United Kingdom Environmental Mutagen Society appointed a Sub-Committee to determine the minimal professional criteria that should be applied to mutagenicity testing in order to meet the requirements of UK authorities. The tests recommended in the 'Guidelines for Testing of Chemicals for Mutagenicity' which was published by the Department of Health and Social Security (DHSS, 1981) formed the initial basis of the first volume which dealt with the most commonly used mutagenicity tests (UKEMS, 1983). A second volume (UKEMS, 1984), which also had to take account of other published guidelines, addressed a series of supplementary tests.

Very few of the chapters in these first two volumes adequately tackled the statistical aspects, either in terms of experimental design, or in terms of data analysis. As many guidelines were employing phrases like 'Data should be analysed using appropriate statistical methods' the UKEMS Sub-Committee decided that Part III of their reports should address the statistical evaluation of mutagenicity test data. This report therefore attempts to do that and, where appropriate, to highlight the statistical implications of experimental design. The topics covered include bacterial and mammalian cell colony and fluctuation assays, *in vitro* and *in vivo* chromosomal aberration tests, sister chromatid exchange tests, *Drosophila* and dominant lethal assays.

2 TERMS OF REFERENCE

The terms of reference of the Sub-Committee were to assess the various statistical approaches available for their suitability in evaluating data from the most widely used mutagenicity tests, such that practising genetic toxicologists would be able to better understand what was required of them by regulatory authorities in this respect, and be better advised as to which forms of analysis were preferred, and why. Specifically for each of the test types, the following items were to be considered:

2.1 How to determine the suitability of the data obtained from an assay for fitting a distribution; when the data are unsuitable; when and how data should be transformed.

2.2 The types of statistical analyses that can be used with the assay data under consideration; which, if any, factors govern the choice of analysis; an order of preference if several types of analysis may be used.

2.3 Some worked examples using real data to help the reader understand 2.1 and 2.2.

3 STRUCTURE

The Sub-Committee consisted of a Steering Group with the task of assessing and reporting on the papers submitted by a series of individual Working Groups (the exception was the introductory paper which was written by one person).

3.1 Steering Group

The main sections of UKEMS were represented by seven individuals and this group was supplemented by three statisticians used to dealing with genetic toxicology data on a regular basis.

3.2 Working Groups

Eight Working Groups were established and chaired by UKEMS members with relevant expertise. The Working Groups comprised between five and nine members, and each group included at least two statisticians.

4 SAFETY CONSIDERATIONS

The safety of staff involved in the conduct of mutagenicity tests described in this and earlier reports has been a fundamental consideration of the Sub-Committee who wish to emphasise that such staff should be fully trained in techniques for handling hazardous chemicals and should be fully aware of the nature of the hazards by reference to the appropriate handbooks (NCI, 1975; IARC, 1979; MRC, 1981).

5 TIMETABLE

The terms of reference of the Sub-Committee require the provision of information that reflects the current state of knowledge of the field.

Reports I and II (UKEMS, 1983, 1984), dealing with genetic toxicology methods in a rapidly developing field, were therefore completed to strict timetables. It was recognised with this report that, although the format or type of genetic toxicology data presented for statistical analysis may change fairly rapidly with time, the statistical approaches would be likely to change less rapidly. It was also recognised that a familiarisation period was required during which genetic toxicologists and statisticians on Working Groups and the Steering Group learned more of each other's disciplines and languages such that communication could be effective. A rigid timetable was not therefore enforced and the entire project spanned from Summer 1985 to Spring 1988.

6 ACKNOWLEDGEMENTS

On behalf of the Sub-Committee and the Society I would like to express my profound thanks to all of the participants in this project and in particular to the statisticians, many of whom were not UKEMS members yet exhibited immense enthusiasm and commitment. I would also like to thank the UK Health and Safety Executive for their financial and moral support of this project.

7 REFERENCES

DHSS (1981). Guidelines for the Testing of Chemicals for Mutagenicity. Prepared by the Committee on Mutagenicity of Chemicals in Food, Consumer Products and the Environment, Department of Health and Social Security. *Report on Health and Social Subjects, No. 24*, Published by Her Majesty's Stationery Office, London, UK.

IARC (1979). Handling Chemical Carcinogens in the Laboratory; Problems of Safety. *IARC Scientific Publication No. 33*. International Agency for Research on Cancer, Lyon, France.

MRC (1981). *Guidelines for Work with Chemical Carcinogens in Medical Research Council Establishments*, Medical Research Council, London, UK.

NCI (1975). National Cancer Institute Safety Standards for Research Involving Chemical Carcinogens. National Cancer Institute, USA. *DHEW Publication No. (NIH) 76–900*. United States Department of Health, Education and Welfare, USA.

UKEMS (1983). *UKEMS Sub-Committee on Guidelines for Mutagenicity Testing*, Report, Part I, Basic Test Battery. Ed. B.J. Dean, Published by the United Kingdom Environmental Mutagen Society, Swansea, UK.

UKEMS (1984). *UKEMS Sub-Committee on Guidelines for Mutagenicity Testing*, Report, Part II, Supplementary Tests, Mutagens in Food, Mutagens in Body Fluids and Excreta, Nitrosation Products. Ed. B.J. Dean, Published by the United Kingdom Environmental Mutagen Society, Swansea, UK.

David J. Kirkland
Sub-Committee Chairman

LIST OF ABBREVIATIONS

$d.f.$ = degrees of freedom
MS = mean square
F or VR = variance ratio
P = probability
NS = not significant
SCP = sum of cross product
D or d = independent variable, e.g. dose
SS = sum of squares
x = dependent variable, e.g. colony count
EMS = error mean square
R = rank
L_j = sum of ranks to cut point 'j'
$S.D.$ = standard deviation
$S.E.$ = standard error
W = weight
MF = mutant frequency
V or Var = variance
OUA = ouabain
6TG = 6-thioguanine
TFT = trifluorothymidine
CHO = Chinese hamster ovary
SCE = sister chromatid exchanges
$ANOVA$ = analysis of variance
z or Z = standard or normal deviate
PE = polychromatic erythrocyte
MPE = micronucleated polychromatic erythrocyte
MTD = maximum tolerated dose
MI = mitotic index
H_0 = null hypothesis
H_1 = alternative hypothesis
α = type I error or probability of false positive conclusion
β = type II error or probability of a false negative conclusion
$1-\beta$ = the power of a statistical test

1

Statistics and genetic toxicology – setting the scene

D.P. LOVELL

1.1 INTRODUCTION

The aim of this chapter is to explain in non-mathematical terms the strengths and limitations of aspects of the statistical analyses that can be applied to genotoxicity studies. First, to improve the understanding of subsequent chapters and statistical ideas in general; second, to try to remove the mystique surrounding aspects of statistics and thereby improve the dialogue between the individuals involved in conducting and assessing genetic toxicology tests.

There is a certain irony in the need for a book of this type at all. Much of current statistical thought developed from the work of researchers interested in the field of genetics. In particular, Sir Ronald Fisher made an enormous contribution to the field of what can now be termed biometrics. Many of the statistical techniques now in common use trace back to Fisher's studies on genetic variation and mutation in a range of species. In fact, one of Fisher's major contributions, the development of the analysis of variance, resulted from his unification in 1918 of the previously independent ideas and warring factions of Mendelian Geneticists and Biometricians (Fisher, 1918).

Fisher was, at the the same time, a geneticist, mathematician, statistician and experimentalist. Such expertise in the same person is rare. In genetic toxicology there is now a division of labour with specialists from different disciplines collaborating. This increased specialism has to some extent broken the statistical links with pioneers in the field.

1.2 WHAT IS STATISTICS FOR?

There are different schools of thought on whether statistical methods should be aimed at testing hypotheses (is the response of the treated group greater than that in the control group?) or estimation (how big is the difference and how confident are we of the difference?). Increasingly, the importance of estimation is being recognised and biomedical statisticians are insisting that statistical practices should reflect this (Gardner & Altman, 1986; Bulpitt, 1987).

An alternative approach is to use statistical methods for exploring and summarising the data. The move away from formal hypothesis testing and significance levels has developed over the last few years following on from the work of John Tukey (1977) in the area called Exploratory Data Analysis. Chatfield (1985) has suggested another somewhat similar pragmatic approach to data analysis called Introductory Data Analysis (IDA).

The increasing use of statistics as a method to explore data sets and to provide an improved summary of the results of an experiment whether by significance tests or exploratory data analysis is more relevant to how genotoxicity assays are actually assessed. The judicious use of statistics as an aid to determining the implications of results from genotoxicity assays is more important than an over-reliance on hypothesis testing methodology.

At times, discussions of statistical philosophy may confirm some biologists' prejudices about how statistics is irrelevant to the work they are doing. It is easy for the non-statistician to become confused and dispirited at the apparent intricacy of statistics and to dismiss its terminology and formality. However, statistics is a complex intellectual activity with roots in mathematics, logic and philosophy as well as the more mundane practice of data processing.

Statistical formulae therefore use mathematical symbols as a concise and precise method of specifying complex ideas. An example which is widely used in other chapters is the symbol $\sum$ (the Greek capital letter sigma) to designate the sum of a set of values.

The full symbol $\sum_{i=1}^{n} X_i$ is a succinct way of saying add all n values of a set of numbers designated $X_1, X_2, \ldots, X_n$. In fact, few of the practical statistical methods described in this book require mathematical skills beyond addition, subtraction, division and multiplication. Most methods are now either programmable for a computer or already exist as programs in software packages (see Appendix 2).

1.3 IS THERE A CORRECT STATISTICAL ANALYSIS?

Non-statisticians may well demand the 'best' approach or the 'most suitable' test. They become confused by the apparent competing claims for different approaches or the seemingly academic disputes over the various methods. Such a response is understandable. Concepts in statistics are still controversial issues raising fundamental questions. There is no single correct statistical method, but, instead, there are different schools of thought with alternative approaches.

An example of the continuing debate is how to analyse a 2×2 table such as the comparison between the proportion responding in control and treated groups – a common design in genetic toxicology. It is over 50 years since Frank Yates first described an analysis for this problem and yet the approach is still contentious. In 1984, on the fiftieth anniversary of Yates' original paper, the Royal Statistical Society held a meeting to commemorate its publication (Yates, 1984). Yates' lecture and the ensuing comments and correspondence takes up nearly 40 pages of the Royal Statistical Society's Journal!

There are certainly correct analyses of the various experimental designs used but there is probably no single correct analysis. In the case of the assays discussed in this book, it is possible to suggest statistical methods that work, i.e. give sensible answers. When more than one test gives sensible answers it may be difficult to decide which, if any, is 'correct'. Alternative analyses may well exist and the use of these methods should not necessarily be discouraged. One good reason is that often little is known of the comparative performance of different statistical tests on real data.

Another reason for not choosing a single 'correct' analysis is that many of the apparently different approaches are often the same statistical analysis in another guise. For instance, the tables of χ^2, t, F and the normal deviate (z) (see list of abbreviations for a description of these terms) found in the back of most introductory statistical text books are interrelated. All the other tables can, in fact, be shown to be special cases of the general F table. It is important, therefore, not to select one method and reject another simply because its terminology or formulation is different. For example, χ^2 tests on proportions can be expressed as tests based upon the normal deviate (z) while the two sample t-test is equivalent to a one-way analysis of variance with two groups.

Many statistical ideas can be related to an underlying general statistical model. An analogy is a continent covered by water with just the mountain tops appearing above the water as islands. Statistical tests like the chi-square and the analysis of variance which may appear independent are in fact inter-connected. The unification of statistical methods and models is an area of considerable intellectual effort which is now meeting with some success (McCullough & Nelder, 1983).

Statistical methods which are recommended in this book should not be taken to be '*the method*' to the exclusion of all other approaches. The methods included in this book may even be idiosynchratic or special cases of a more relevant general technique chosen because of familiarity with a particular approach. Computers and statistical packages are becoming more powerful and more widely accessible so that new statistical methods which will be more appropriate could replace today's optimum approaches.

It is appropriate to warn, however, that even with computers incorrect analyses can be carried out. There is no limit to people's ingenuity in the misuse of statistical packages. It is not possible to warn against all such types of errors in detail. However, some of the incorrect approaches based upon misunderstandings about statistical ideas are included in the discussion on each genotoxicity assay. The main way to guard against the use of incorrect methods is to maintain a familiarity with the data being analysed and by a healthy scepticism of computer printouts!

The aim of the individual chapters is to provide the statistical methods necessary for the genetic toxicologist to analyse the results of assays. Most of the methods are in fact straightforward and relatively easy to apply but difficulties may arise. If so, stop, and consult a statistician. The explanations in the chapters should provide sufficient information for a dialogue to take place between the genetic toxicologist and the statistician.

1.4 EXPERIMENTAL DESIGN

The experimental designs used in genotoxicity testing are not very different from those used in other areas of toxicology. The organisms may differ – bacteria, mice or mammalian cells; the end-points may vary – point mutations, chromosomal abberrations or abnormal foetuses – but the underlying experimental concepts are similar. A standard assay usually includes a negative control and a treated or experimental group as well as the inclusion of a positive control (where a treatment known to produce an effect is administered). There may also be a dose-response experiment to investigate the relationship of the end-point to variation in the doses. The experimental designs developed for genotoxicity assays do not differ appreciably from those developed by Finney and co-workers in the 1950s and 1960s for biological assays in general (Finney, 1978). Certainly, some of the statistical methods used in biological assay could be applied to the analysis of data from mutation experiments.

Statistical methods have been developed to try to detect effects of treatments above the 'noise' of biological variability. Variation can be present at any of a number of levels depending upon the mutagenicity assay; for instance, between cells, between plates or cultures, between animals, and between observers. Experimental designs are methods for taking this variation into account, estimating it and minimising its effect on the factors or the treatments which are the main purpose of the study. Choosing sample sizes, types of samples, doses, all form part of determining the experimental design. The presence or absence of variability at different levels of the design can influence how acceptable it is to pool data from different units such as plates or animals. The choice of the *experimental unit* in the analysis is important because variation between cultures in a chromosome aberration study or between males in a dominant lethal study may affect the significance testing of hypotheses.

In many of the more sophisticated statistical analyses the experimental design is a fundamental aspect of the analysis determining both the nature and the power of the statistical tests to be used. Identification and incorporation of potential sources of variation into the experimental design as a constituent part of the study can subsequently provide considerable insight into how successfully the objectives of the experiment were achieved.

1.5 STATISTICAL ISSUES RELEVANT TO ALL MUTAGENICITY ASSAYS

Some of the experimental designs permit a number of possible methods for analysis. Examples of the range of methods actually in use will be found in the subsequent chapters; however, a number of statistical issues (Table 1.1) need to be discussed which are relevant to an appreciation of statistical methods in general.

1.5.1 Probability and hypothesis testing

Probability is a very practical concept understandable to any gambler, whilst at the same time being a deeply philosophical statistical concept. An understanding of how often events occur in the 'long run' is a sufficient start for the interpretation of statistical tests. Some events are common although their actual occurrence at a particular point is uncertain; others are rare. Probability is measured on a scale from 0 to 1 where 0 means an event will never happen and 1 means that it is certain to.

One aspect of a statistical analysis is to provide some indication of whether the results obtained were compatible with some hypothesised effect. In fact a hypothesis, called the *null hypothesis*, is set up that suggests, for instance, that a particular treatment has no effect. The null hypothesis is deliberately created to be challenged and possibly rejected in favour of an *alternative hypothesis* that the treatment does have some effect. (Underlying this somewhat artificial formulation of the purpose of an experiment is

Table 1.1. *Statistical issues of relevance to all mutagenicity assays*

1.5.1	*Probability and hypothesis testing* Null hypothesis, alternative hypothesis, test statistics (F, t, χ^2, z), critical values, significance levels. Statistical significance is *not* a measure of the size of an effect.
1.5.2	*One- or two-sided significance tests* One- or two-sided test, one- or two-tailed statistical distributions, statistical tables of distributions.
1.5.3	*The power of a statistical test* Neyman-Pearson theory of hypothesis testing, false positives, false negatives. Prediction, sensitivity, specificity, accuracy, prevalence, Type I (α) error, Type II (β) error, power ($1-\beta$).
1.5.4	*Biological and statistical significance* Hypothesis testing, estimation, exploration of data. Statistical tests on positive controls. 'Statistical sanctification', high dose effects, thresholds, monotonic dose-response, statistical significance and/or biological importance, correlation.
1.5.5	*What to do with borderline significance levels* Weak effects or Type I errors, repetition of borderline experiments, pooling results, combination of probability values, sample size and power.
1.5.6	*Replication and randomisation* Replication of and randomisation of experimental units.
1.5.7	*Distribution of data* Parametric tests, normal distributions, 'robustness'.
1.5.8	*Transformations* Counts, Poisson, square root transformation; proportions, binomial, arcsine or angular transformation; logarithmic transformation, scales, back-transformation.
1.5.9	*Parametric v non-parametric tests* Parametric methods: Analysis of variance, t-test, linear regression, assume normal distributions, equal variances, independent data. Non-parametric methods, distribution free tests, relative rankings.
1.5.10	*Outliers* Artefacts, blunders, technical errors.
1.5.11	*Testing comparisons between means* Planned, *a priori*; unplanned, *a posteriori*; multiple comparisons, orthogonal comparisons. Dunnett's test, Bonferroni correction.

the concept that while a hypothesis can be disproved it can never be proved.)

The decision whether or not to reject a null hypothesis is based upon comparing the test statistic (e.g. F, t, χ^2, etc.) obtained from the statistical test with tables of *critical values* for that particular test. The critical values are those associated with particular *significance levels.*

These significance levels provide a probabilistic statement that the results observed occurred by chance alone when the null hypothesis is true. For historical reasons based upon the availability of suitable tables a number of critical values have become common. These are based upon probabilities of a by-chance-alone effect of one in 20, one in 100 or one in a 1000 experiments. These are sometimes expressed as percentages or as probabilities. Various ways of representing these effects are shown in Table 1.2.

There is no compulsion to be restricted to these critical values and increasingly statistical software packages are printing the actual probability levels associated with the statistical test. Many statisticians prefer to report these values arguing that is is misleading to draw arbitrary lines such that a probability level of 0.051 is non-significant while a level of 0.049 is significant.

It is important to appreciate that the significance level is not an absolute measure of the size of the difference between, say, two group means. It is a function of the experimental design and the statistical tests used.

Rare events which have small but finite probabilities do occur. Therefore an extremely small probability may represent a real effect of a treatment but it might also mean that this is a rare chance effect. This is an important consideration when many comparisons are being made. If, for instance, all the possible comparisons between 100 means are carried out, there would be 4950 comparisons. If many comparisons are made, or many experiments are carried out, rare chance effects will eventually be detected. In fact, the choice of specific significance levels actually defines the number of events that occur by chance. The problem of multiple comparisons will be discussed in more detail below.

1.5.2 One- or two-sided significance tests

The non-statistician is often confused by the problem of whether a particular statistical method involves a one- or two-sided significance test. The agonies of choosing which test can be avoided if certain key ideas are thought through. The core of a statistical test is the Null Hypothesis that a treatment has no effect. At the same time an alternative hypothesis is also proposed. This can either be that the experimenter is interested in a difference between the control and the treated (the difference can be in either direction and the statistical test is then two-sided) or that the experimenter is only interested in treatments which alter the incidence in one direction by increasing, for instance, the incidence of chromosomal aberrations (in this case the test is one-sided).

The actual choice of critical values for determining significance levels for testing the null hypothesis is determined from statistical tables and needs some care as the tables appear in many forms. The one-sided test is a less conservative test than the two-sided in that smaller effects will be

Table 1.2. *Different methods of showing statistically significant results*

Statistically significant			Expressed as				
percentages	by proportions	ratios	probability less than	or	probability between	Expressed as stars[a]	Expressed as words[b]
5%	0.05	1/20	$P<0.05$		$0.01<P<0.05$	*	Significant
1%	0.01	1/100	$P<0.01$		$0.001<P<0.01$	**	Highly significant
0.1%	0.001	1/1000	$P<0.001$		$P<0.001$	***	Very highly significant

Notes:
[a] Sokal & Rohlf (1969) p.169.
[b] Sprent (1981) p.44.

considered to be significant. A one-sided test is therefore appropriate when the direction of the effect is of interest: the treatment may or may not increase the damage or number of mutations compared with the control and whether or not it reduces the incidence below the control level is not of concern. (It is, however, important to check that incidences are not reduced in treated groups as such a finding might indicate problems with the experimental procedures such as high dose toxicity. This is an example of using statistics for exploring patterns in the data rather than for formal and precise significance testing.)

In general, one-sided statistical tests should be used in genotoxicity assays. (Conceivably, in a research project, an unknown compound might be either a mutagen or an anti-mutagen protecting against genetic damage, and experiments designed to determine which.) Care is needed to ensure that the correct sided test is in fact carried out especially when using statistical packages. Some tests such as the analysis of variance and testing for linear trends are routinely reported by many statistical packages as two-sided tests. Such tests will underestimate the statistical significance of any treatment related effects.

A further complication is that it is not always apparent from a statistical table whether the associated probability levels are for one- or two-sided tests. For instance, the table of critical values for the t distributions given by Snedecor & Cochran (1967, Table A4) are for two-sided tests while those given by Winer (1971, Table C2) are for a one-sided test. Also, tables such as those of the F and χ^2 distributions give the probabilities associated with *one tail* of the distribution but which are appropriate for a *two-sided* test. Figure 1.1 shows the appropriate tails of the distribution to use for either a one- or two-sided test. Mather (1964, Section 18, p. 46) provides a more technical but helpful description of the relationship between the various distributions.

The symbol χ^2 can be confusing because it seems to appear in two forms: either the Greek letter chi (χ) or the Roman X. This is because the Greek and Roman letters are used in general to distinguish between a theoretical value and an observed value. The *theoretical* chi-square distribution is denoted by χ^2 while the *calculated* chi-square statistic is referred to as X^2. A similar distinction is made between the theoretical mean and standard deviation of a population called μ and σ (Greek letters mu and sigma) and the observed values obtained in an experiment designated $\overline{X}$ and s.

1.5.3 The power of a statistical test

The formulation of a statistical analysis in the form of a test of a hypothesis (called the Neyman-Pearson Theory of hypothesis testing after

Fig. 1.1. Frequency curves of z, t, χ^2 and F distributions showing the respective tails of the distributions used for taking critical values for two- and one-sided tests at the $P < 0.05$ criterion.

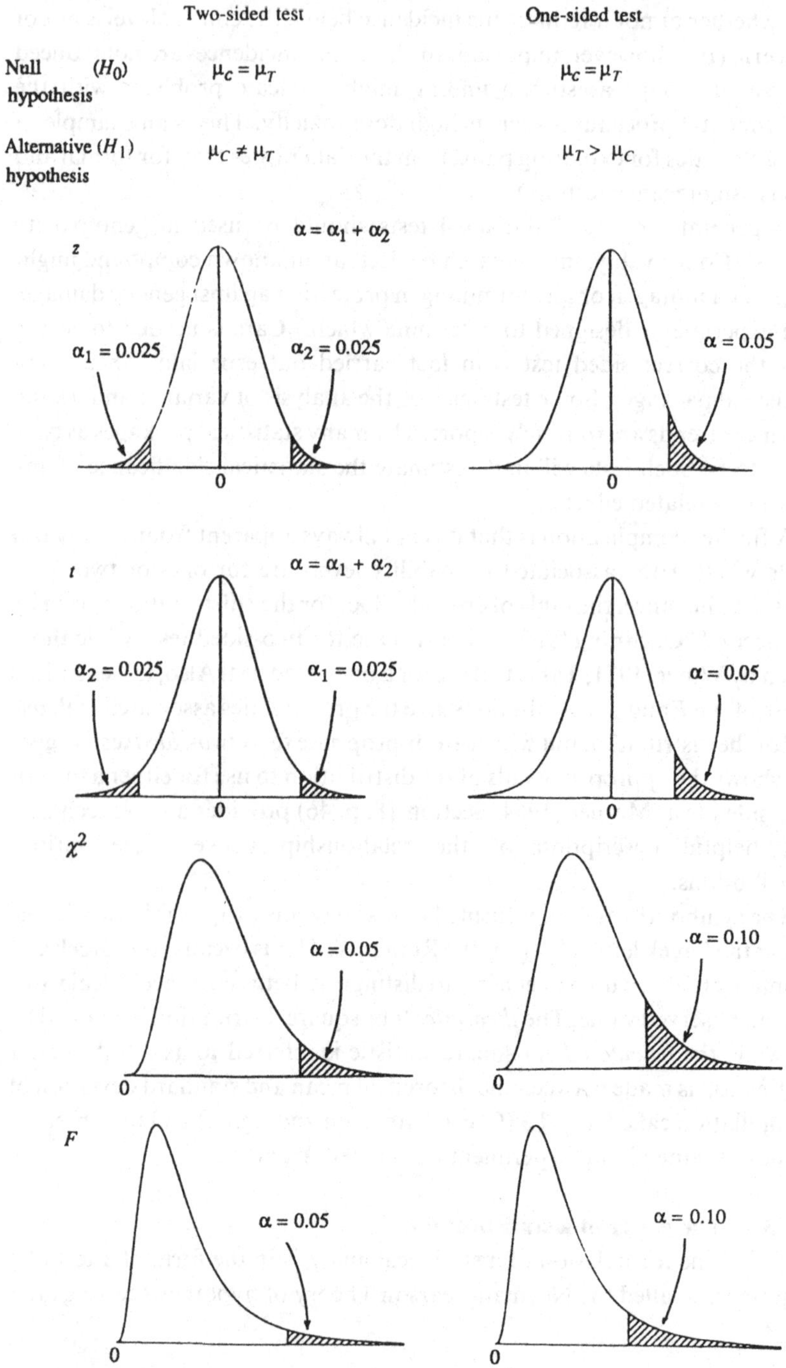

	Two-sided test	One-sided test
Null (H_0) hypothesis	$\mu_C = \mu_T$	$\mu_C = \mu_T$
Alternative (H_1) hypothesis	$\mu_C \neq \mu_T$	$\mu_T > \mu_C$

Box 1.1. *Possible outcomes of prediction of carcinogenicity from a mutagenicity assay*

		Carcinogenicity	
		+	−
Mutagenicity	+	True positive *a*	False positive *b*
	−	False negative *c*	True negative *d*

Notes:
$N = a + b + c + d$
Derived terms
Sensitivity = proportion of carcinogens positive in mutagenicity test $a/(a+c)$
Specificity = proportion of non-carcinogens negative in mutagenicity test $d/(b+d)$
Positive predictive value = proportion of positive results which are carcinogens $a/(a+b)$
Negative predictive value = proportion of negative results which are non-carcinogens $d/(c+d)$
Accuracy/validity = proportion of correct results with all tests $(a+d)/N$
Prevalence – number of carcinogens in total chemicals tested $(a+c)/N$

the originators of the methods) can result in two types of mistakes: first claiming to find a difference which, in fact, does not exist and secondly, failing to find a real difference. Such errors are similar to those found in diagnostic tests for medical conditions or in the use of genotoxicity tests to predict the potential of a chemical to produce a carcinogenic hazard. Box 1.1 illustrates the possible outcomes of such studies. The genotoxicity assay may have either a positive or negative result, the chemical may or may not be a carcinogen. In two cases the genotoxicity assay could correctly predict the potential carcinogenicity of the chemical; the true positive and true negative. In two other cases the prediction is wrong: in the false negative the genotoxicity assay was negative but the chemical was a carcinogen; while in the false positive, the genotoxicity assay was positive but the chemical was a non-carcinogen. The false positive and false negative rates (together with the prevalence or ratio of carcinogens to non-carcinogens) can be used to provide various measures for assessing the performance of an assay (Cooper, Saracci & Cole, 1979). These terms such as sensitivity, specificity and positive or negative predictivity are estimated as shown in Box 1.1 and are derived from concepts developed in diagnostic tests.

Table 1.3. *Outcomes of testing a null hypothesis from a mutagenicity assay*

	Null hypothesis (H_0)	
	False	True
Reject null hypothesis	Correct decision	Type I error α (also known as – Error of the first kind or Producer's risk)
Accept null hypothesis	Type II error β (also known as – Error of the second kind or Consumer's risk)	Correct decision

A similar type of table can be drawn up for the results of testing a null hypothesis (e.g. Table 1.3). The main difference is that the decision to accept or reject the hypothesis under test is based upon the probability of the observed result of an experiment occurring by chance. Figure 1.2. illustrates the two types of error that can arise. Curve 1 shows the distribution of differences between the observed control mean ($\overline{X}_C$) and the observed treated mean ($\overline{X}_T$) on the assumption of no difference between the true means (μ_C and μ_T). The critical value is shown by the vertical straight line. This is chosen so that a specified proportion of the differences will be considered significant (or the null hypothesis rejected) where in fact there is no difference. (This is the significance level, or the Type I error.) Conversely, if the null hypothesis is not true with μ_T greater ($>$) than μ_C the distribution of the differences is shown by Curve 2. Differences above the critical value will be considered significant but a proportion β will be below the critical value and considered non-significant. This proportion β is referred to as the Type II error. The value $1-\beta$ is a measure of the power of a statistical test or the ability of the test to detect a true difference of a certain size. Any statistical test is a trade-off between the two error rates. The power of a test can be increased by large sample sizes, improved experimental design and choosing appropriate statistical tests. Previous knowledge of the variability of the material under test should provide some guide to the size of experiment necessary to detect certain events. Alternatively, certain experimental designs can be shown to have little chance of detecting effects of a specified size. Note again that the level of significance is not a measure of the size of a difference but is a function of the experimental design and the statistical methods applied.

Although superficially similar Box 1.1 and Table 1.3 are based upon different concepts. A source of confusion to the genetic toxicologists is that terms like false positive, sensitivity and specificity are sometimes applied to statistical hypothesis testing while the proportion of false positive and negative results in Box 1.1 are sometimes designated as the α and β rates.

In Table 1.3 the errors are a consequence of variation due to sampling; in Box 1.1 false positives and negatives may be real effects. For instance an animal carcinogen may not be genotoxic in the particular assay or a chemical that is genotoxic may not reach a target site in the animal to cause cancer.

1.5.4 Biological and statistical significance

The concepts behind statistical methods of hypothesis testing and determining statistical significance outlined in the previous two sections have often caused confusion to biologists. Similarly, to single out hypothesis testing as the main purpose of statistical analysis at the expense of the other functions of estimation and exploration of data can also hinder understanding.

Fig. 1.2. Diagrammatic representation of the two types of errors (α and β) arising from a hypothesis test. Curve 1 is the frequency distribution of the difference between the observed control ($\overline{X}_C$) and treated ($\overline{X}_T$) means assuming no difference (the null hypothesis $\mu_C = \mu_T$). Curve 2 is the frequency distribution if the difference when treated is greater than the control (alternative hypothesis $\mu_T > \mu_C$). The ordinate is the frequency and the abscissa is a measure of the difference between the treated and control means.

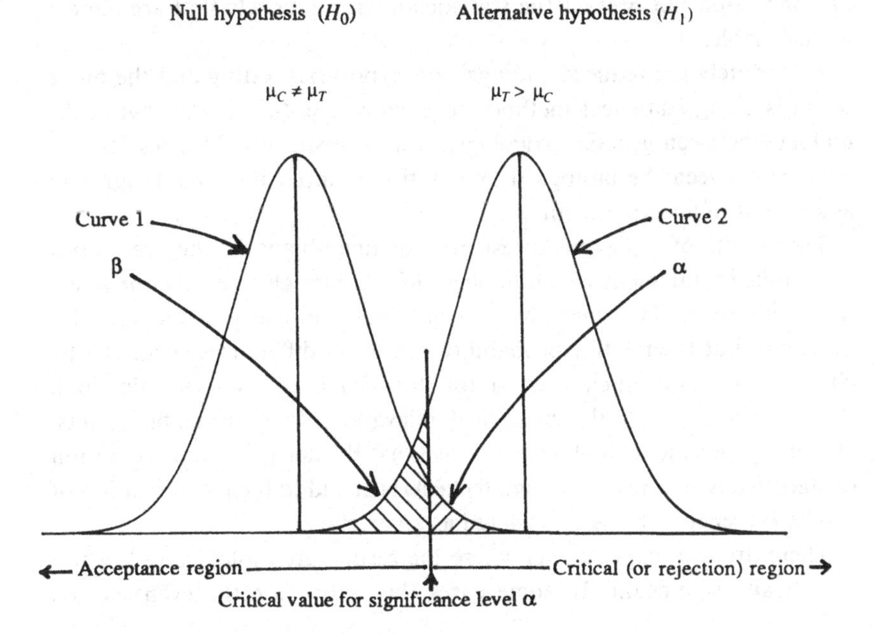

As a consequence statisticians are aware of a response to their work which has become a cliché: 'it may be statistically significant but it is not biologically significant!'. The nature of the hypothesis testing procedure also suggests that it should provide a decision making process so that various decision criteria are suggested in genetic toxicology protocols and phrases such as 'unambiguous positive response', 'without resort to statistics', 'absence of a *clear* dose-response relationship', 'consistent with', 'biologically sensible relationships' are used. Definitions of what these terms mean are not included. 'Experience and sound scientific judgement' may be needed to interpret the results but that includes the interpretation of statistical analyses rather than its avoidance as is implied by some of these phrases.

Any reluctance to use statistical methods probably results from unfortunate previous experience. The 'I don't need statistics to tell me that' approach is partly a response to the increasing requirement for biological journals to seek, in John Tukey's words, 'statistical sanctification' of the data submitted to them. The use of unnecessary statistical tests to show the effects of a positive control where a large and predictable response has occurred is another example. However, in many other studies statistical methods can detect such effects using fewer resources or may alternatively detect effects and trends not immediately apparent. Unfortunately some researchers still design experiments to produce a qualitative answer but find the results are not clear cut, turn to statistics for confirmation of their preconception but present the statistician with data sets that are almost unanalysable.

Fortunately the reduced emphasis on hypothesis testing and the move towards using statistical methods to explore the data is improving the dialogue between genetic toxicologists and statisticians. This results in a balance between the biological and statistical input into the design and assessment of an experiment.

The results of a statistical test are not unambiguous, they rely on a probabilistic statement which implies *either* a rare chance effect *or* a real treatment effect. The significance level is not a measure of the size of a difference but the relative probability of certain differences occurring by chance alone. The judicious use of the statistical conclusions should aid in the interpretation of the biological relevance of experimental results. Hopefully, genetic toxicologists will feel that the detection and estimation of statistically significant but small effects is an aid to the interpretation of results rather than being considered as an artefact.

There are also experiments where the results are biologically but not statistically significant. In some cases this is because the event occurs extremely rarely (such as chromosomal exchanges in cytogenetic studies) and methods can be developed for the inclusion of such rare phenomena into some form of statistical analysis. In other cases it is unfortunately the result of poor experimental design. Most of the assays discussed in the subsequent chapters have some basic protocol and it should be feasible to estimate the size of effects that are detectable statistically and then decide what action to take. Designing experiments which have little chance of detecting effects as statistically significant, but which are considered biologically meaningful, is a culpable waste of resources.

The apparent contradiction between biological and statistical significance is illustrated by high dose effects of some test compounds. Many experiments show increasing responses with increasing dose – a monotonic response. In some cases the response is only found at high doses with no apparent responses at the lower doses. The identification of a threshold, if any, is often statistically difficult while a statistical analysis can give little insight into the relevance of a high dose effect. Biological knowledge is needed to argue whether such a finding represents a true genotoxic effect or an artefact based upon an unnatural overloading of the homeostasis of the biological system.

Similarly the monotonic dose-response may not occur because of a reduced response at high doses. Often this is a toxic response to high doses of the chemical. While statistical analyses can identify such findings, biological knowledge is necessary for the interpretation.

The word 'significant', in effect, has two meanings: a closely defined term as used by statisticians and a more general term as used by biologists. A useful distinction could perhaps be made by talking of statistical significance or biological importance. (A similar confusion exists with the term 'correlation' which has a precise definition in statistics but is used by non-statisticians in a more general form as a synonym for 'mutually related'.)

1.5.5 What to do with borderline significance levels

The interpretation of statistical analyses is likely to be difficult if the results are close to one of the accepted statistical significance levels such as $P=0.05$. A strict interpretation of a statistical test would be to call all results with significance levels greater than $P=0.05$ non-significant (this is often seen in reports with either *NS* or $P>0.05$ written against the comparison). Others, however, are less precise in their description of how to interpret findings at these arbitrary, but now conventional, borderlines. Cox (1958), for instance, gave the following practical meaning to a result statistically significant at or near the $P=0.05$ level but not near the $P=0.01$

level: 'data give good evidence that the true contrast (effect) is not zero'. Clearly, this is a point where both statistical and biological expertise becomes important.

In a study using appropriate statistical methods borderline statistical effects may represent either a weak effect or a Type I error. Distinguishing between the two will always be difficult. The distinction between small effects and borderline significance is important. It is, for instance, possible to show a highly significant trend with increasing dose by a linear trend test (Box 1.2) but for this slope to be very gradual. Such an effect would almost certainly be real but may be of marginal biological importance. The detection of significant effects should be reported but interpretation needs to be based upon expert knowledge of this system. 'Biologically sensible relationships' may well be subjective and the source of expert debate, but the use of isolated probability statements will be of little help. However, data from one experiment may help in designing subsequent studies.

Repeating an experiment with the same design will probably produce another borderline result especially if there is a genuine but small effect. Pooling the results from a series of experiments is possible in certain cases but the statistical methods depend upon whether the experiments form part of a series or are carried out under different conditions. Issues such as potential biases in the choice of data for inclusion in such tests are important. In general, expert statistical advice should be sought.

Sokal & Rohlf (1969, p.621), however, present a simple method developed by Fisher for estimating the probabilities associated with a series of independent, but not necessarily similar, experiments testing the same hypothesis. The natural logarithms ($\log_e$) of the probabilities associated with each experiment are summed, multiplied by -2 and compared with an χ^2 with degrees of freedom equal to twice the number of experiments. Clearly, whether or not to include experimental results can make a considerable difference to the interpretation of these combinations. This method can only be used when all the probability values are known which is a point in favour of the reporting of actual probability values instead of using the asterisk and *NS* notation.

In practice, decisions about the biological significance of detecting chemicals with small or weak effects is beyond the scope of the statistical analysis of a study. Increasing the size or power of a study may confirm weak effects as real, but then makes it likely that even smaller effects will be detected as borderline significant. Genetic toxicologists need to be prepared to report results where interpretation is not necessarily straightforward. Quantifying the effects detected appears a more intellectually defensible position than trying to ignore them. Considerations of the power of an

Box 1.2. *Illustration of a statistically significant result whose biological importance may only be defined in light of data from other investigations*

Data are imaginary colony counts from an Ames test-type experiment conducted at three doses and a control level.
At each dose three replicate plates are scored.
The raw data are:

Dose	Colony counts			Mean counts
0	31	36	29	32
1	38	41	47	42
2	46	53	57	52
3	70	63	56	63

The data are analysed by linear regression for illustration (see Chapter 2 for a full discussion of appropriate statistical tests for Ames test-type data). The analysis of variance table below shows a highly significant linear regression ($P < 0.001$) or increasing incidence of colonies with increasing dose when compared with the variability between plates within a single dose (referred to as residual in the table below).

Source	*d.f.*	*MS*	*F*	*P*
Linear regression	1	1591.4	55.84	<0.001***
Heterogeneity	2	0.4	<1	*NS*
Residual	8	28.5		
Total	11			

An understanding of this table is not strictly necessary (Snedecor & Cochran, 1967 p.456 provided a full description of the statistical methods used) but the data provide an example of an effect which is very highly significant and likely to be real but where none of the mean counts at any dose levels are more than twice the control value of 32. This doubling value is used by some as a criterion for a positive effect in Ames test experiments.

experimental design may be a helpful concept to include as part of the statistical assessment of data in such studies.

1.5.6 Replication and randomisation

Two important concepts underlying many experimental designs are the replication and randomisation of experimental units. Many designs require an estimate of variability between units to compare with the variability found between treatments. Replication of the experimental units within a particular treatment provides this. An underlying assumption behind many statistical analyses is that the data come from a random sample. Randomisation of the experimental units into the treatment groups is an important procedure both to minimise the consequence of any pre-existing variability in the sample and to reduce the risk of bias resulting from uncontrolled factors during the conduct of the experiment. Care needs to be taken that randomisation is carried out so that each unit has an equal chance of being exposed to a particular condition during the experiment. Failure to randomise can produce spurious results. Randomisation is of particular importance in animal studies but should equally be considered for non-animal experiments. A reduction in the risk of false results may, in the end, adequately compensate for the extra complexity in the methodology of a study.

1.5.7 Distribution of data

Statistical tests in many cases assume certain underlying mathematical models with the data being selected from a population distributed in particular ways.

Within each of the particular types of distribution a specific distribution can be completely defined by the parameters which describe its shape (in the case of a normal distribution, the parameters are its mean and standard deviation). This results in the name 'parametric' for statistical methods based upon assumptions about the data's distributions. If the data are a good fit to these distributions, powerful and efficient statistical tests are produced. Even when some of the underlying assumptions are not exactly met, the conclusions drawn from the analyses will not be seriously wrong. The methods are said to be 'robust'.

Defining when the analyses are no longer appropriate is not a simple task but depends to some extent on the personal viewpoint of the analyst. Methods for checking various assumptions behind the statistical methods used are discussed at relevant points in the following chapters.

Two alternative approaches are available if the data do not conform to the assumptions underlying a particular analysis. These are to try various transformations or to use non-parametric (or, as they are also called, distribution free) analyses.

1.5.8 Transformations

Data are transformed to try to ensure that the assumptions underlying parametric methods of analysis are more nearly met. Measurements in the form of counts are often distributed as a Poisson rather than as a normal distribution. In the Poisson distribution the mean is equal to the variance. A suitable transformation is the use of square roots of the original data (or, when these include a number of zero or small counts, the square root of the original values plus a half i.e. $\sqrt{(X+0.5)}$).

A transformation called the arcsine (or angular) transformation is applied to data which are proportions and therefore distributed as a binomial. Most sets of statistical tables provide values of the common transformations, or the transformation can be included in the computer program used for the analysis.

Fig. 1.3. Diagrammatic representation of the logarithmic transformation of data. The upper distribution is a log normal distribution of a biological variable measured on a scale *X*. The lower distribution is the normal distribution produced by a logarithmic transformation of the data. Three representative data points are illustrated joined to their transformed values on the log *X* scale. The ordinate is frequency. The respective boxes illustrate how the transformation produces equal within group variability. Note that the mean of log transformed data is not equal to the logarithm of the mean of the untransformed data.

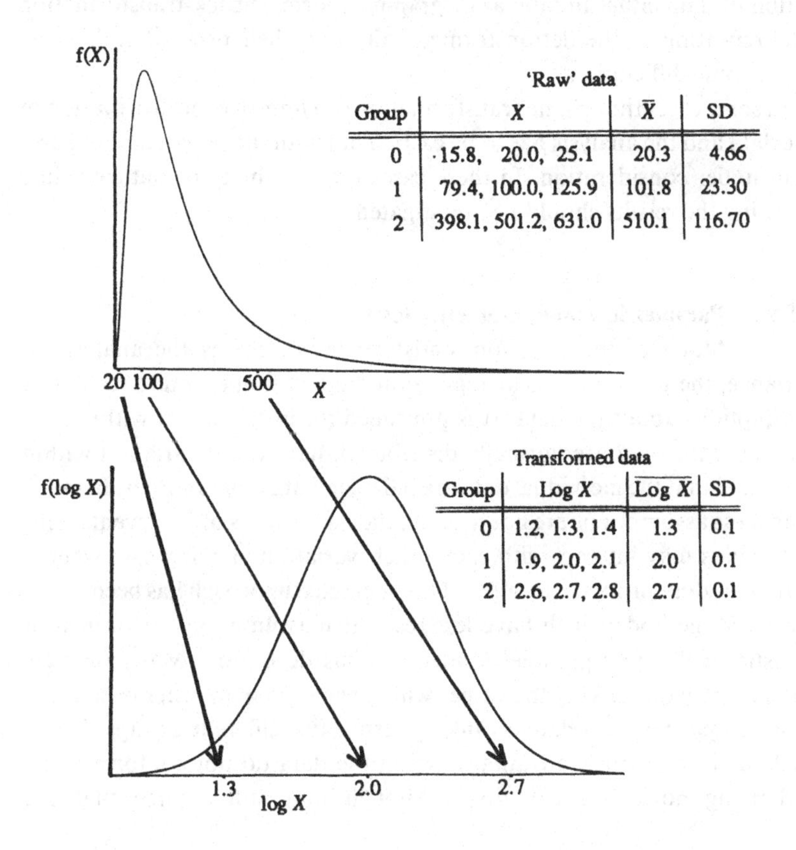

'Raw' data

Group	X	$\bar{X}$	SD
0	15.8, 20.0, 25.1	20.3	4.66
1	79.4, 100.0, 125.9	101.8	23.30
2	398.1, 501.2, 631.0	510.1	116.70

Transformed data

Group	Log X	$\overline{\text{Log } X}$	SD
0	1.2, 1.3, 1.4	1.3	0.1
1	1.9, 2.0, 2.1	2.0	0.1
2	2.6, 2.7, 2.8	2.7	0.1

The third commonly used transformation is the logarithmic. This is used when the mean is positively correlated with the variance. An example of the effect of the use of this transformation is given in Fig. 1.3. The logarithmic transformation is often a successful method of ensuring that biological values are approximately normally distributed.

Others can be used, many of which form a set of related transformations with varying degrees of severity. The choice of a suitable one is, to a large extent, a result of experience in handling data.

Many biologists, though, have considerable doubts about the validity of transforming data. It is associated with 'manipulating' or 'massaging' the data, almost as if it was an illicit act! These worries have probably been one of the spurs to the adoption of non-parametric methods. Certainly there is some difficulty in the concept of measuring data on different scales, though few have difficulty in understanding pH as a measure of the negative logarithmic concentration of hydrogen ions. Measurements are made using scales which are convenient but there is often no reason why these units have to have an underlying biological importance. However, representation of data either in tabular or graphical form as back-transformation (the reporting of the detransformed values on their original scale) may cause some difficulties.

In some cases though, no transformation can improve the distribution of the data and the analysis has to be carried out with the possibility of biases kept under consideration. In these cases the possibility of outlier values affecting the results should be investigated.

1.5.9 Parametric v non-parametric tests

Many of the common statistical tests such as the analysis of variance, the *t*-test and linear regression are based upon quite restrictive assumptions about the data. It is presumed for instance that within each treatment the data are normally distributed, have equal variation within doses and that the individual data are independent of one another. In many tests these assumptions are not true; the data are counts of rare events; cells or foetuses may interact with each other; variation may increase when a particular dose produces an effect. One suggested approach has been to use statistical methods which have less restrictive assumptions. A branch of statistics called non-parametric methods has developed. Many of these methods rely on ranking the values within each group in order of size and then comparing the relative rankings from the different groups. These methods have distinct advantages when the data do not conform to an underlying normal distribution (although they still assume that the individual data are independent of one another) and they then provide more accurate significance tests. They have also been recommended because they are more easily understood by the non-statistician, can, in some cases, be nearly as powerful as the parametric tests even when the data are normally distributed, and provide a more general approach to the analysis.

Both approaches have their proponents. Certainly, if one wishes to ensure that the Type I error rate (rejecting the null hypothesis when in fact it is true) is accurate, then the choice of an appropriate test is paramount. However, if estimation of effects and data exploration are considered equally important, obtaining a Type I error that is absolutely accurate becomes less relevant.

Non-parametric tests have the disadvantages that they are not as well known, can require some computing expertise, are not easily generalised to more complex designs and, because they use ranks rather than the individual values, are, in general, less powerful than the parametric alternatives when the data are normally distributed.

Nevertheless, in most simple instances there is a non-parametric analysis complementing the parametric approach and they are acceptable tests in many cases. Computer programs are now available which can deal with the large amount of computing needed for comparing ranks.

1.5.10 Outliers

In all experiments there is some variability in response but some values are more extreme. In some cases this can result in the assumptions inherent in parametric analyses not being met. Such outliers are a source of difficulty in the analysis of data. Do they, for instance, represent part of the true underlying variation in the cells, plates or animals, or are they artefacts or rogue values? Can they be excluded, especially if they appear to be having a disproportionate effect on the results? However, both the scientist and the regulator may be concerned at the exclusion of data from the analysis. How extreme does the outlier have to be for exclusion? There are no easy answers. If something is known about the datum to attribute confidently its variation to some technical accident or mistake in measurement (sometimes defined as a 'blunder') or if it is biologically unusual (and here there is room for debate among experts), it can be excluded provided this is noted. If it is unusual only because it is large or small then exclusion is not so easy to justify. Various statistical methods exist for trying to identify and then remove potential outliers (Barnett & Lewis, 1978). They rely on a series of assumptions and none provides clear and dependable guidelines. A commonsense approach is to analyse the data with and without the outlier(s) and see if the interpretation of the data is dramatically altered. As in many areas, the biologist's judgement is used (and of course may be questioned by other biologists).

1.5.11 Testing comparisons between means

Comparison between a series of means can be either planned as part of the experimental design, or only be suggested once the results of a study have been obtained. The statistical methods used depend upon whether the comparisons were planned or unplanned.

Planned comparisons (also called *a priori* tests) are hypotheses included either implicitly or explicitly as part of the experimental design. An example is a test for a linear trend in a dose-response study. In some studies a set of independent hypotheses can be tested with the property that each hypothesis has a Type I error (the risk of falsely rejecting the null hypothesis) determined by the significance level. Such sets of hypotheses are called orthogonal comparison. Not all sets of *a priori* comparisons are, however, orthogonal.

In the case of unplanned comparisons (also called *a posteriori* tests), if all possible comparisons between the means are carried out, then the actual Type I error will be greater than the significance levels used for testing the hypothesis (i.e. the risk of falsely rejecting the null hypothesis will be increased above, for example, the 5% level).

A considerable number of different methods for making such multiple comparisons have been suggested to avoid the risk of inflating the Type I error. A comprehensive review of methods is given by Dunnett & Goldsmith (1981). A consequence of the use of multiple comparison methods is that a comparison is tested in effect at a low significance level, so that there is also a reduced chance of detecting a real effect as significant.

Two commonly used methods are the Dunnett's test and the Bonferroni correction. Dunnett's test was developed to compare a series of different treatments against a negative control. The method ensures that a treatment will be falsely called 'different from the control', on average, only once in 20 experiments (if the 5% significance level is used) when none of the treatment groups is different from the control.

The Bonferroni correction provides a significance level adjusted to take into account the number of possible comparisons that could have been made. If comparisons are to be made between pairs of a set of five means, the critical significance levels for each comparison will be adjusted by dividing the levels by the number of possible comparisons (in this case ten). Therefore, if comparisons were being tested at the 5% significance level, the actual test would be made using the 0.5% (5%/10) level.

Clearly the interpretation put on statistical analyses in which multiple comparison methods are used may differ from those when they are not. Care is needed in assessing results when such methods have been used, as an interpretation based solely on probability levels might suggest some compounds were less potent mutagens.

1.6 THE USE OF HISTORICAL CONTROLS AND BAYESIAN METHODS

Most statistical tests are directed at analysing the data derived from a single experiment. However, laboratories have often collected considerable control or background data on the materials or organisms they use. It is possible under certain circumstances to incorporate such information into a formal statistical analysis rather than for it to be used as part of a subjective interpretation of the study. Such methods do have to be used with considerable care to ensure that the various control groups are compatible, and probably should only be used after consultation with a professional statistician.

A branch of statistical thought which is likely to become increasingly relevant in the assessment of genotoxicity data is the use of Bayesian statistics. These are named after Thomas Bayes, an eighteenth-century clergyman, who developed an alternative view of probability to that described earlier. Bayesian methods assume that there is some prior knowledge of the likely outcome of an experiment either based upon probabilities observed in previous studies or from a 'degree of belief' in the outcome. Methods involving Bayesian ideas have been developed for the inclusion of historical control data. Another use of Bayesian methods has been in attempts to develop batteries of short-term genotoxicity tests to identify potential carcinogens. In this case the prior probabilities, as they are referred to, are based upon the performance of the short-term tests when used on a set of known carcinogens and non-carcinogens. The use of statistics to investigate problems of prediction are somewhat similar to diagnostic tests in medicine, and the issues arising are somewhat different from those involved in the analysis of a single experiment.

1.7 THE USE OF DECISION RULES

The results of statistical analysis may either be a probabalistic statement of how likely is a difference between e.g. the means of a control and treated group to be a chance finding, or form a confidence interval of the estimate of the difference. They are not necessarily a definition *per se* of a positive or negative experimental result.

An experiment may be analysable by many possible statistical methods. In some cases it is possible to produce contradictory results. For instance, a significant overall dose-response relationship can be found but individual comparisons of the dose levels against the controls are non-significant. One approach to the interpretation of these sorts of data is the development of decision rules which formalise the interpretation of the statistical tests. Galloway *et al.* (1985) have developed and applied such rules to the analysis of data from *in vitro* assays for sister-chromatid exchange and chromosomal aberration. They categorised chemicals on the basis of results from two sets of statistical analyses as positive, negative, weak positive, or questionable positive, as well as positive but with no dose-related effect. The choice of the criteria for each category was based upon experience of the use of the assay. Such decision rules could be developed for a variety of assays but would need some assessment to see if the decisions being produced were successful at identifying potential hazards. They do, however, show how the statistical and biological interpretation can be included to provide a 'result' to a study.

1.8 FUTURE DEVELOPMENTS IN STATISTICS RELEVANT TO GENETIC TOXICOLOGY

Statistical research is an active field. Methods of statistical analysis appropriate at present may be improved or superseded. In particular the increasing power and availability of micro- or personal computers increases the range of methods available to the genetic toxicologist. Software for carrying out statistical analyses is likely to continue to develop. Some of the non-parametric approaches relying upon considerable calculations of relative rankings, and other methods which involve intensive computational procedures, become more feasible as computing power increases.

Further research is likely on the relative performance of different statistical approaches to the analysis of the type of data produced in genetic toxicology experiments. This will involve the simulation of representative data using a computer. These methods, called Monte Carlo methods, will also explore different experimental designs and allocation of resources within the design. Statistical methods incorporating more background knowledge, such as historical controls, will continue to develop, as will methods for the inclusion of both statistical and biological expertise into decision rules.

1.9 CONCLUSIONS

Genetic toxicology studies need statistical analyses to provide some estimate of whether the results obtained were likely to occur by chance, or as a result of exposure to a genotoxic agent. Such assessment is not necessarily easy, but an argument can be made for the use of the most efficient and informative statistical techniques to allow for full examination of the data.

Experimental design and statistical analysis are both fundamental parts of a genetic toxicology study. The users of such analyses must be aware of their responsibility to understand and interpret the statistical methods involved as part of the overall conduct of such assays.

1.10 REFERENCES

Barnett, V. & Lewis, T. (1978). *Outliers in Statistical Data.* Wiley, Chichester.

Bulpitt, C.J. (1987). Statistical analysis-confidence intervals. *The Lancet*, **i**, 494–7.

Chatfield, C. (1985). The initial examination of data (with Discussion). *Journal of the Royal Statistical Society* **A 148**, 214–53.

Cooper II, J.A., Saracci, R. & Cole, P. (1979). Describing the validity of carcinogen screening tests. *British Journal of Cancer*, **39**, 87–9.

Cox, D.R. (1958). *Planning of Experiments.* Wiley, New York.

Dunnett, C. & Goldsmith, C. (1981). When and how to do multiple comparisons. Ch. 16. In: *Statistics in the Pharmaceutical Industry*, Ed. C.R. Buncher & Jia-Yeong Tsay. Dekker, New York, pp. 397–433.

Finney, D.J. (1978). *Statistical Method in Biological Assay* 3rd edn. Griffin, London.

Fisher, R.A. (1918). The correlations between relatives on the supposition of Mendelian inheritance. *Transactions of the Royal Society of Edinburgh*, **53**, 329–433.

Galloway, S.M., Bloom, A.D., Resnick, M., Margolin, B.H., Nakamura, F., Archer, P. & Zeiger, E. (1985). Development of a standard protocol for *in vitro* cytogenetic testing with Chinese hamster ovary cells: comparison of results from 22 compounds in two laboratories. *Environmental Mutagenesis*, **7**, 1–51.

Gardner, M.J. & Altman, D.G. (1986). Statistics in medicine: confidence intervals rather than *P* values: estimation rather than hypothesis testing. *British Medical Journal*, **292**, 746–50.

Mather, K. (1964). *Statistical Analysis in Biology*, 5th edn. Methuen, London.

McCullough, P. & Nelder, J.A. (1983). *Generalized Linear Models*, Chapman & Hall, London.

Snedecor, G.W. & Cochran, W.G. (1967). *Statistical Methods*, 6th Edn. Iowa University Press, Iowa.

Sokal, R.R. & Rohlf, F.J. (1969). *Biometry* Freeman, San Francisco.

Sprent, P. (1981). *Quick Statistics. An Introduction to Non-Parametric Methods.* Penguin, Middlesex.

Tukey, J.W. (1977). *Exploratory Data Analysis.* Addison-Wesley, Massachusetts.

Winer, B.J. (1971). *Statistical Principles in Experimental Design* 2nd edition. McGraw-Hill, New York.

Yates, F. (1984). Tests of significance for 2 × 2 contingency tables (with Discussion) *Journal of the Royal Statistical Society* **A 147**, 426–63.

2

Analysis of data from microbial colony assays

G.A.T. MAHON M.H.L. GREEN
B. MIDDLETON I. de G. MITCHELL
W.D. ROBINSON D.J. TWEATS (Group Leader)

2.1 INTRODUCTION

Venitt *et al.* (1983) have reviewed all aspects of the performance of bacterial mutation assays including the interpretation of results. Parry *et al.* (1984) have similarly reviewed the mutation assays involving yeast cultures. The main emphasis in the present chapter is on the factors affecting the statistical analysis, the choice of method to be used and the performance of the analysis. Statistical analysis of data from such assays may be divided, rather arbitrarily, into two aspects. On the one hand, the biology of the microbial mutation systems may be considered and the statistical implications derived. On the other hand, one may begin from statistical principles and explore their field of application in the light of the biology. The two approaches lead to complementary sets of conclusions.

Venitt *et al.* (1983) and Parry *et al.* (1984) discuss a number of topics closely related to statistical analysis, in particular, recording and storage of data, presentation of results and the minimum data to be presented, interpretation of data in terms of positive and negative, and dealing with ambiguous results. These topics are to some extent also developed in the present chapter.

These authors also point out that a wide range of different statistical approaches has been advocated. This Working Group has been able to recommend some approaches in preference to others. However, this should not be taken to imply either that other methods are always invalid or useless, or that interpretation based on other methods is necessarily unsound.

As most experience has been obtained with reversion assays such as the Ames test, many of the examples given refer to such assays. However, the statistical methods outlined and the conclusions drawn are equally applicable to other 'mutational' or 'conversion' assays, which entail the analysis of colony counts.

2.2. BIOLOGICAL ASPECTS

2.2.1 Colony-forming microbial assays

There are several ways of performing microbial assays for which the endpoint is discrete colonies. These have been discussed previously (Venitt *et al.*, 1983; Parry *et al.*, 1984). A brief summary of the two major types of technique is given below as a framework for statistical discussion.

2.2.1.1 *Plate incorporation assays*

The *Salmonella*/mammalian microsome assay or Ames assay (Ames *et al.*, 1975; Maron & Ames, 1984; Kier *et al.*, 1986) is the most widely used mutagenicity test procedure. Test compounds are incubated with specially constructed strains of *Salmonella typhimurium*, selected for sensitivity and specificity in reversion from histidine requirement (auxotrophy) to histidine non-requirement (prototrophy). This method can also be adapted to other microbial species with different auxotrophic requirements, for example, the *E. coli* assay.

Essentially the assay is carried out as follow: 10^7 to 10^8 cells from late log phase cultures of *Salmonella* are added to molten overlays containing a traced of histidine, test compound and, when required, liver homogenate plus co-factors. After mixing, the overlays are poured over the surface of solid minimal agar plates and these are incubated at 37 °C for 2 or 3 days. The trace of histidine in the top agar plus the nutrients from the bottom agar allow limited bacterial growth. This is required for the fixation of the optimum number of pre-mutagenic lesions in the DNA and results in a thin lawn of histidine-requiring bacteria. Only the small number of bacteria which have reverted to histidine prototrophy will continue to divide to form visible discrete colonies, each originating from a single revertant cell.

The usual format of this and similar assays consists of a positive control, one or two negative controls and five concentrations of agent; all are plated at least in duplicate, often in triplicate or more. Agent-treatments are spaced at two- to five-fold concentration intervals. For each compound one such test is performed in the absence of microsomal activation and one in the presence of activation. This whole assay is normally repeated (once).

2.2.1.2 *Treat and plate assay*

Treat and plate assays, in which the agent is incubated with the microbial cells prior to plating, are to be preferred to Ames assays in three situations. Certain agents or their metabolites (e.g. dimethylnitrosamine) are very unstable or short-lived and thus their activity may be detected more readily in a liquid culture system than in agar plates, where diffusion may be a problem. Likewise agents which are only mutagenic at toxic levels may be more easily detected when survival is taken into account. Lastly in many treat and plate assays, the treatment and selective systems are separated so the possibility of artefacts (e.g. growth stimulation) due to their interaction is eliminated.

In selecting the inoculum size for the treatment phase of treat and plate assays, there are several important, and to some extent conflicting, factors to take into account. To ensure an adequately sensitive assay the number of exposed cells should provide a sufficiently large target to detect the small number of mutants induced by the agent at the locus under study. However, the inoculum must not contain too many pre-existing mutants, produced during growth (and treatment), or the few induced mutants will be obscured by the many pre-existing mutants and much sensitivity will be lost. Finally, the number of spontaneous mutants, at the plating stage, must not be so small that there are likely to be major variations in the numbers of spontaneous mutants distributed between control and treated cultures. A more detailed discussion of the importance of spontaneous mutants to test sensitivity is given in Section 2.2.2.

There are several protocols for treat and plate assays. One (Yahagi *et al.*, 1977) is a variation of the Ames assay and survival as such is not measured. In this assay the agent is separated from the selective system during an initial incubation period in liquid and then a plate incorporation assay is carried out as described in Section 2.2.1.1. Thus this system is identical to an Ames assay from a statistical aspect, but solves the problem of diffusion of both short-lived metabolites and soluble enzymes from the activation system.

Treatment can take place in non-nutrient medium that will not allow cell division (Green & Muriel, 1976) or in culture medium (Parry *et al.*, 1984). Treatment may be followed by a period of growth in nutrient medium in the absence of agent (Mitchell *et al.*, 1980; Thompson & Melampy, 1981). Cells are then plated out on selective agar. Cells are also plated on non-selective (nutrient) agar to measure survival. It is only in the absence of the growth of unmutated cells on selective plates that direct comparisons can be made between the numbers of mutants per survivor at each treatment. If unmutated cells are allowed to grow, e.g. by adding a trace of relevant amino acid to strains with auxotrophic markers, then the number of mutant colonies obtained will remain the same as the control values even when survival has been reduced by several orders of magnitude for the reasons given below (Section 2.2.2). Thus if the number of mutants per survivor are calculated for a non-mutagenic but toxic compound a false positive result will be obtained (Green & Muriel, 1976). This danger of miscalculation has been known for over 30 years but it is still frequently ignored in the literature.

The assay format is more variable than for Ames tests. Single to triple cultures are used per agent- (control-) concentration, with three to five agent dose-levels. There is a positive control and one or two negative controls. Each culture (the 'unit' of test) is usually plated out on a minimum of three selective and three non-selective plates. For each compound one such assay is performed without microsomal activation and usually one with activation. Assays are normally repeated (once).

The protocols used for the treat and plate assay are often similar to those of mammalian cell gene mutation assays. The analysis of data from such mammalian cell assays is discussed in Chapter 3 and the statistical methods recommended in that chapter may also be recommended for the bacterial treat and plate assay.

2.2.2 Spontaneous mutation and sensitivity of assays

In detecting induced mutation, it is important to remember that spontaneous mutations arising before and after the period of treatment will reduce the sensitivity of the test. The culture used for the experiment will contain mutant bacteria, descendants of those that mutated during the growth of the culture. Depending on whether mutations have occurred early or late in the growth of the culture, there may be many or few mutants. The extreme variability in the number of mutants between replicate cultures was the basis of the original demonstration by Luria & Delbrück (1943) of fluctuation in bacterial cultures and it is important to realise that cultures for experiments grown from small inocula are precisely those that will show the greatest relative variability in mutant frequency. Moreover, each mutation that occurs before the bacteria are finally plated may give rise to several or many mutant colonies (Green, 1981). In the Plate Incorporation protocol, induced mutations will arise after the bacteria are plated and each mutation will give rise to only one colony. Even in a treat and plate protocol, induced mutants will have at most a limited opportunity for division before plating. Spontaneous mutations occurring before the period of treatment and before plating will thus cause a disproportionate loss of sensitivity of the test.

A further problem with pre-existing mutants occurs with many Treat and Plate protocols where aliquots of bacteria are dispensed and allowed a period of growth in liquid medium during and after treatment. When the bacteria are dispensed, pre-existing mutants will be distributed according to the Poisson distribution, but as they continue to divide the variation in mutant frequency between cultures will be maintained and the distribution of numbers of mutants will become hyperpoisson.

An illusory solution to these difficulties would be to reduce the number of bacteria treated to a level where pre-existing mutants were rare. Such a practice is particularly dangerous because it may not be apparent from the data presented. In the Plate Incorporation protocol and many Treat and Plate protocols a period of growth after treatment is allowed in order for induced mutations to express. In the case of histidine reversions this is achieved by placing a limiting amount of histidine in the plating medium. Non-mutant bacteria grow to exhaust the histidine. It is reiterated that the number of *his*$^+$ revertants will be largely independent of the number of bacteria plated and almost entirely determined by the amount of histidine added. If the inoculum is reduced this will probably not be apparent from the numbers of mutants on the spontaneous plates. If, in a Plate Incorporation protocol, the agent tested has a short half-life, reducing the inoculum will reduce the number of cells treated without this being apparent from the data. A similar situation will occur with a Treat and Plate method where growth of non-mutant cells is limited in this way.

2.2.3 Critical factors which affect reproducibility between tests

There are numerous aspects of assays which will affect reproducibility of results within and between laboratories. Problems can arise with the physical and chemical properties of test compounds (e.g. volatility, solubility and stability). Microbial strains can be highly specific in what they detect because of specificity of the DNA changes required, 'genetic drift', strain specific metabolism, or differences in spontaneous frequency. The activity and specificity of mammalian microsomes is very dependent on the donor species used, method of induction, method of preparation, method of storage and quantity used. The protocol is critical, viz. agar- versus liquid-based assays, open versus closed systems, choice of solvent and numerous other assay variables.

These considerations will affect assay to assay variation. They are important in the assessment of individual assay function by comparison with the historic data base in a laboratory, or in the literature. However, a different set of factors governs the ability to determine whether a treatment has produced a statistically significant increase in mutation relative to the control within an assay.

2.2.4 Critical factors which affect the determination of statistical significance

Some of these factors have already been reviewed (Parry *et al.* 1984). Essentially they can be divided into factors underlying the basis of the analyses (i.e. the distribution) and factors affecting the sensitivity (i.e. experimental design).

2.2.4.1 *Distributional considerations*

(a) Operations such as weighing and pipetting cannot be performed precisely and the quantities weighed or pipetted are subject to small, random errors. These errors would be additional to errors from other sources and should be similar across assays.

(b) The transfer of discrete units (microbial cells) from one vessel to another or to petri plates is also liable to error. Such errors would be expected to follow a Poisson distribution but may diverge from this ideal if microbial cells tend to stick together in clumps, although good experimental technique should minimise this possibility. Thus the variance of the number of bacteria transferred could be increased. The increase will become critical only when low numbers of cells are being handled.

(c) Variations associated with the mutation process itself theoretically follow a Poisson distribution of genetic events (Luria & Delbrück, 1943). However, there are factors which may result in greater variation (Campbell, 1974). For example, the Poisson distribution follows from the assumption that equal numbers of cells are exposed; variation in toxicity between samples or significant cell division during treatment but prior to plating will lead to greater than Poisson variation.

Difficulties in transfer of bacteria and in the mutation process tend to become more acute, in terms of divergence from the Poisson distribution, at toxic levels of the test agent. Many of the mutagens found in routine screening of industrial compounds, drugs, etc, are weakly mutagenic and only active in assays at, or near, toxic levels. Thus many of the methods suitable for the statistical analysis of potent standard mutagens may be less suitable when applied to weaker agents.

2.2.4.2 *Experimental design, protocols and sensitivity*

The factors in experimental design which affect the sensitivity of statistical analyses may appear obvious but their inter-relationships can be more complex.

(a) Inoculum size is important because the greater the number of induced mutant colonies the greater the statistical sensitivity. However, there is a trade-off between microbial sample size and the number of replicates. Samples of a size which lead to colony counts of zero should be avoided and, ideally, each plate should contain at least five mutant colonies to give adequate statistical sensitivity and permit effective transformation of the data.

(b) Replication varies in importance depending on the form of statistical analysis to be used. For parametric methods (where the variance depends on the mean and the form of the dependence is to be estimated from the data) and for many non-parametric methods, as many replicates as possible (at least four replicates) should be included. Where the dependence of the variance on the mean is fixed through a specific assumption, three replicates are adequate, but it would not be possible to verify the assumption from individual experiments.

(c) The number of treatments, provided there are at least two, is less important than inoculum size or replication unless trend analysis is to be used. Generally, trend analysis is more relevant to determining dose relationships (once statistical significance has been established) than to the determination of significance itself. Furthermore, the more treatment levels there are, the more likely it is that one will show up by chance as statistically significant and thus a larger allowance for this factor is necessary in the statistical analysis, i.e. in pairwise comparisons with control, increasing the number of treatment levels can decrease the statistical sensitivity of the assay, particularly where agents are active only over a narrow dose range, as is often the case for weak mutagens active only at toxic levels.

(d) Dose interval is important because it is desirable that the region of mutagenic activity (if any) is covered. Doses outside this region (apart from the controls) are of limited value. If doses are spaced too close together, no discernible trend can be observed, whilst if too far apart the region of activity may be missed. Most agents, if they are mutagenic, show greatest activity near the toxic limit for the test. Two- to three-fold concentration intervals have been found to work adequately, going downwards from the toxicity or solubility limit.

(e) Repeat testing is an effective way of reducing statistical errors (false positives or false negatives) in terms of return on effort. However, if the data from repeat tests do not agree, further tests will be needed.

2.2.5 Regulatory requirements and determination of statistical significance

The requirements or suggestions in many regulations for five dose levels at up to a five-fold concentration interval using duplicate or triplicate plates may not be the optimum design for the determination of statistical significance; more replicates and fewer dose levels may be preferable. Many data are generated according to the usual experimental design to satisfy regulatory requirements and so the analysis of such data will be discussed in detail. As improvements in techniques are often encouraged by regulatory agencies, they will be suggested as appropriate. Regulatory implications of the Ames assay in a wide context have been reviewed by Jackson & Pertel (1986).

2.2.6 Interpretation of statistical significance

Prior to investigating methods for determining statistical significance, it is important to decide on their interpretation. A positive (one-sided) significance test will show that if the control and treated sample were the same, then it is unlikely (at a given probability) that they would have yielded data as disparate as those observed, with the inference that the test agent produces values which exceed those of the control. However, this does *not* necessarily mean that the compound is mutagenic. The following alternative interpretations are possible:

(a) The result is a chance statistical finding, a false positive.
(b) There is an assay artefact (e.g. growth stimulation in the Ames assay).
(c) There is a mutagenic impurity. In this case the differences tested for statistical significance cannot be attributable to the compound alone.

Statistical analysis can show whether a difference is likely to be due to chance, but it is the responsibility of the scientist to decide whether the difference is a real effect (i.e. can be attributable to the treatment alone) and to determine whether artefacts or impurities need consideration. While the determination of statistical significance is a critical aid to interpretation, the scientist must decide whether the result has relevance *in vivo*.

2.3 STATISTICAL ASPECTS

2.3.1 General considerations

The conclusion from an experiment depends not only on the activity (if any) of the test agent, but also on the design and protocol of the experiment, the amount and nature of sampling variation, and the properties of the statistical test applied. Several of the topics considered in the following paragraphs are discussed from a different viewpoint and often in detail in Chapter 1 of this volume. A framework for the statistical evaluation of genotoxicity data in general has been presented by Mitchell (1987). Recent work on the analysis of Ames assay data in particular in the US National Toxicology Program has been reviewed by Margolin (1985).

2.3.1.1 *Sampling variation*

The starting point of any discussion of sampling variation in the Ames assay or other microbial assays must be the Poisson distribution. If the number of bacteria per treated or control plate is the same and if each bacterium has an equal chance of mutating and yielding a revertant colony, then the number of revertant colonies will have a Poisson distribution. The relative variability of a set of data may be measured and tested by means of the χ^2 statistic (see Box 2.1 and Section 2.3.3.1 for an example). The χ^2 value divided by its degrees of freedom gives the m statistic. The m statistic may also be defined as the sample variance divided by the mean; however, the calculation of m from χ^2 gives a single m value for an experiment with several subgroups. Several authors have reported data more variable than would have been expected from the Poisson distribution (Margolin *et al.*, 1981; Bernstein *et al.*, 1982), while others have reported data consistent with that distribution (Stead *et al.*, 1981). Variation within and between experiments in a large collaborative study has been analysed by Margolin *et al.* (1984).

The sources of variation in the colony counts within a dose level have been discussed in Section 2.2.4.1. The relative importance of these sources may be deduced. If 100 μl of bacterial suspension contains 10^6 cells, and if the number of cells follows a Poisson distribution, then its standard deviation will be 10^3, and the coefficient of variation of the number of cells will be $10^3/10^6$ or 0.1%. The coefficient of variation due to pipetting has been found empirically to be of the order of 1% to 5%. If the number of revertant colonies per plate is about 100 then the coefficient of variation in that number due to Poisson sampling will be of the order of 10%. It is difficult to know how much variation is inevitably caused by other factors. However, some sources of variation are under the control of the experimenter and, with good technique, variation from these sources should be low compared to Poisson variation in the number of colonies. Thus within-experiment variation close to that implied by the Poisson distribution is a reasonable goal for the experimenter.

The data reported by Margolin *et al.* (1981) are of particular interest because of the unusual degree of replication used. Five sets of colony counts

Box 2.1. *Worked example:* χ^2 *and* m *statistics*

Dose	Counts (x)			$\sum x$	$\sum x^2$	SS_x	$\overline{x}$	$d.f.$	χ^2
0	129	128	119	376	47186	60.67	125.33	2	0.484
0.5	128	116	121	365	44481	72.67	121.67	2	0.597
1.7	127	142	112	381	48837	450.00	127.00	2	3.543
3.8	165	152	159	476	75610	84.67	158.67	2	0.534
5.0	178	149	163	490	80454	420.67	163.33	2	2.576
								10	7.734

Notes:
$m = 7.734/10 = 0.77$.

with 20-fold replication for strain TA100 were produced. Four sets of data were water controls and one set DMSO. The least variable sets of data had a variance close to that implied by the Poisson distribution but the most variable set had a variance nearly four times greater. Taken together, the variance of all sets of data exceeded the Poisson expectation by a factor of 2.4. The question arises as to whether these data, especially the more variable ones, are typical of what can be achieved in practice.

Ten sets of control data with strain TA100 each with 20-fold replication were produced for consideration by this Working Group by Glaxo Group Research Ltd. The data and corresponding statistics are shown in Box 2.2. The least variable data show a level of variation consistent with the Poisson distribution. The variation of the most variable set exceeded that level by a factor of 2.34. Overall, the variance of the data exceeded the Poisson expectation by a factor of only 1.27. Thus, it appears that with TA100 data can be produced for which the variance is close to that of the Poisson distribution although somewhat more variable data should be expected. However, data that are four times more variable cannot be regarded as normal.

Experience in Beecham Pharmaceuticals has shown variability of data close to the Poisson expectation for strains TA98, TA1535 and TA1538 with or without microsomes and for TA100 without microsomes. Data were about two-fold more variable for TA97 with or without microsomes and for TA100 with microsomes.

The m statistic provides a measure of the relative variability of a set of data, and m will be about 1 if the data have a Poisson distribution. If, for a particular set of data, m is less than 1, then the Poisson distribution may be assumed, since there is no biological or experimental reason to expect the true variance to be *less* than Poisson. If m is between 1 and, say, twice the average historical value in a laboratory then one of the methods of analysis which allows variation greater than the Poisson distribution should be used (Section 2.3.2). If m is greater than twice the historical average then the experiment should be repeated and the first experiment discarded. On the one hand, a negative result with very variable data would carry little weight since the extra variation might have obscured a weak positive result. On the other hand, a positive result with such data might be due to the extraneous source of the variation rather than the activity of the test agent. It is unclear just what limit on variability should be set above which an experiment should be repeated, though it seems clear that some such limit based on experience should be decided.

2.3.1.2 *Types of statistical error*

A central element in the role of statistical analysis is its use in decision making. One might wish to decide if a test agent is or is not effective in elevating the number of revertant colonies on a plate. After carrying out an experiment, one would calculate a statistic and then accept or reject the hypothesis that the agent is ineffective on the basis of the value of the statistic.

It is clear that an incorrect conclusion may be reached in one of two ways: a true hypothesis may be rejected (a Type I error), or a false hypothesis may be accepted (a Type II error). The probabilities of the types of error under a given set of circumstances are denoted α and β respectively

		Decision	
		Accept	Reject
Hypothesis	True	$1-\alpha$	α
	False	β	$1-\beta$

Thus α is the probability of reaching a false positive conclusion while β is the probability of a false negative conclusion. Furthermore, α is the significance level and is most usually set at 5% while $1-\beta$ is termed the power of the test.

The two probabilities, α and β, are more or less under the control of the investigator and the seriousness of errors of either type may vary with circumstances. The common 5% significance level may be regarded as a convention of scientific reporting; a less stringent level (say 10%) would lead to too many spurious effects being reported while a more stringent level (say 1%) would be wasteful of effort. The emphasis is on avoiding reporting false positive results.

From the point of view of screening it may be more important to avoid false negative results. Thus protocols are laid down which specify the minimum scale of experiments. However, it is not clear that the usual scale of experiments has been chosen so that the magnitude of response of a particular size will be detected with a particular power. It is usual to test if treatment values are significantly above controls. However, fewer false negatives might be obtained by testing whether treatment values are significantly below some ceiling. The development of this approach is beyond the scope of the Working Group. However, the idea of such a threshold test has been considered in defining a negative result in the mouse specific locus assay (Selby & Olson, 1981).

In the context of research, experimental designs and statistical tests may be compared by applying them with a certain significance level and then

Box 2.2. *Ten sets of control data from Strain TA100*

	A int		*A-D*		*B int*		*B-D*		*C*	
	108	109	116	85	107	125	135	102	100	108
	99	100	104	107	134	134	122	113	105	93
	97	100	109	102	116	110	121	108	126	82
	125	107	109	99	109	104	120	108	104	98
	120	108	118	102	131	114	123	115	108	89
	108	107	95	85	124	109	125	106	99	102
	97	113	93	117	138	118	112	134	112	93
	99	108	131	103	115	143	141	112	98	105
	110	95	118	110	105	152	113	99	126	96
	118	106	100	101	87	109	112	106	93	103
$\sum x$		2134		2104		2384		2327		2040
$\sum x^2$		228934		223824		288874		273141		210320
$\bar{x}$		106.70		105.20		119.20		116.35		102.00
SS		1236.20		2483.20		4701.20		2394.55		2240.00
χ^2		11.586		23.605		39.440		20.581		21.961
d.f.		19		19		19		19		19
m		0.610		1.242		2.076		1.083		1.156

	A1 int		*A1-D*		*B1 int*		*B1-D*		*C1*	
	129	122	90	110	118	130	120	127	132	118
	113	126	106	108	144	123	120	110	122	102
	125	108	112	95	130	124	102	110	114	126
	130	152	112	127	114	127	115	125	99	151
	122	119	97	103	141	135	119	116	130	139
	116	102	116	124	131	128	114	105	122	91
	121	115	122	101	121	125	115	119	131	116
	97	103	79	108	134	119	119	119	136	87
	106	109	112	101	112	134	146	140	116	107
	123	115	113	109	127	107	110	110	130	103
$\sum x$		2353		2145		2524		2361		2372
$\sum x^2$		279723		232577		320202		280845		286592
$\bar{x}$		117.65		107.25		126.20		118.05		118.60
SS		2892.55		2525.75		1673.20		2128.95		5272.80
χ^2		24.586		23.550		13.258		18.034		44.459
d.f.		19		19		19		19		19
m		1.294		1.239		0.698		0.949		2.340

Note:
Total $\chi^2 = 241.1$, total *d.f.* = 190, overall $m = 1.269$.

comparing their power. The comparison may be made analytically on the basis of theory or by computer simulation.

2.3.1.3 *Randomisation*

In human and animal experimentation it is usual to assign individuals randomly to treatment or control groups. In microbial work, where the experimental material may have less intrinsic variability, the use of randomisation is not usual. Since randomisation has certain advantages (outlined below) its application to microbial experiments will be presented although its use is optional. A disadvantage of randomisation is that a more complicated experimental routine would increase the likelihood of mistakes.

Randomisation assures the validity of statistical analyses. In the context of the parametric methods, it makes it possible to obtain valid or unbiased estimates of experimental variation, treatment means and differences among them. In the context of the non-parametric methods, randomisation assures that in the absence of a treatment effect the particular realisation of ranks, signs, or other values may validly be regarded as one of a set of all possible realisations, each one being equally likely. In the presence of a treatment effect, randomisation assures that an appropriate feature of the test statistic may validly be identified with the treatment effect.

In practice, randomisation involves several steps. In an experiment with three control plates and three plates at each of four doses, the plates may be numbered as follows:

Dose	Plate number
0	1, 2, 3
1	4, 5, 6
2	7, 8, 9
3	10,11,12
4	13,14,15

A random permutation of the number 1 to 15 may be obtained using a table of random numbers (e.g. Steel & Torrie, 1960, Table A.1). Successive pairs of digits are examined and noted provided that the value of the digits is in the range 01 to 15 and has not already been noted. The following permutation was obtained using the first three columns in Steel & Torrie (1960), p.430:

08,	13,	12,	07,	14,	05,	11,	02,	04,	10,	15,	06,	09,	03,	01
A	*B*	*C*	*D*	*E*	*F*	*G*	*H*	*I*	*J*	*K*	*L*	*M*	*N*	*O*

For successive experiments, different parts of the random number table would be used to find different permutations and hence different dose orders. In relation to the experiment, plates would be prepared, stacked and counted in the randomised order, *A, B, C, . . ., M, N. O.*

Dose	Plate
0	*O, H, N*
1	*I, F, L*
2	*D, A, M*
3	*J, G, C*
4	*B, E, K*

That is, for all operations from the beginning to the end of the experiment, the following dose order would be used:
2, 4, 3, 2, 4, 1, 3, 0, 1, 3, 4, 1, 2, 0, 0
This obviously involves more work than the usual procedure which would begin with the three control plates, then the three plates for the lowest dose, and so on.

2.3.1.4 *Toxicity and the form of the dose-response*

Although a linear dose-response (Fig. 2.1a) is often found in microbial mutation experiments, there may be a progressive falling off in the slope of the dose response so that the response at high doses tends towards a plateau (Fig. 2.1b). Indeed, at high doses, there may be a decline in the frequency of revertant colonies with increasing dose (Fig. 2.1c). This falling off is usually attributed to toxicity and at very high doses there may be a visible thinning of the bacterial lawn indicating toxicity. McGregor *et al.* (1984) have examined the relationship between toxicity and the number of revertant colonies. They conclude that while a decline at high doses may often be due to toxicity this is not necessarily so. With 2, 4-dimethoxyaniline and TA98 + S9, for example, a reduced number of colonies at high doses was observed but no evidence of toxicity was found.

When the dose response is non-linear but with no actual decline in number at high doses (Fig. 2.1b) then there is no particular difficulty in the statistical analysis. There are several possibilities for handling such data, including a transformation of the dose scale. If doses are transformed to logarithms, or square roots, or reciprocals, an essentially linear dose response may be obtained.

When there is a decline in revertant frequency at high doses (Fig. 2.1c) there are several possible approaches to the statistical analysis, none of which is entirely satisfactory:

- the high dose points may be removed; the significance level will be affected; if a Bonferroni correction is applied, much power may be lost:

Fig. 2.1. Forms of dose-response curve: a. Linear response; b. Non-linear response: plateau; c. Non-linear response: decline.

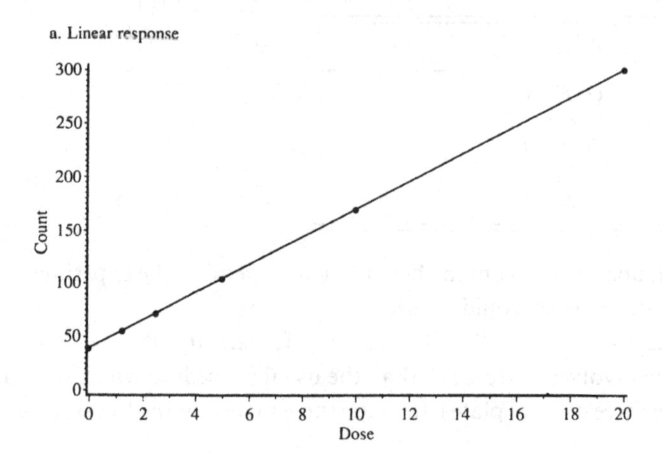

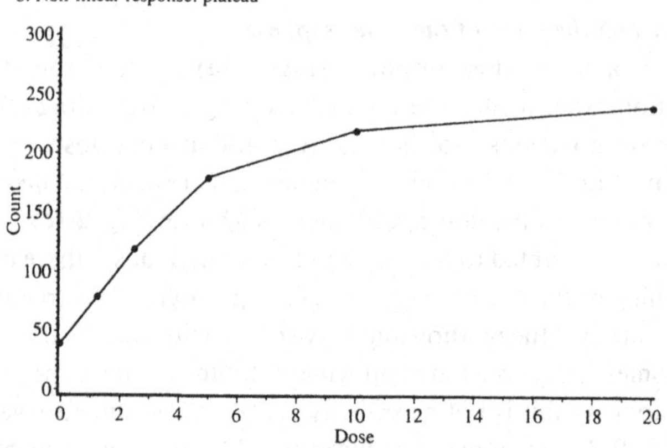

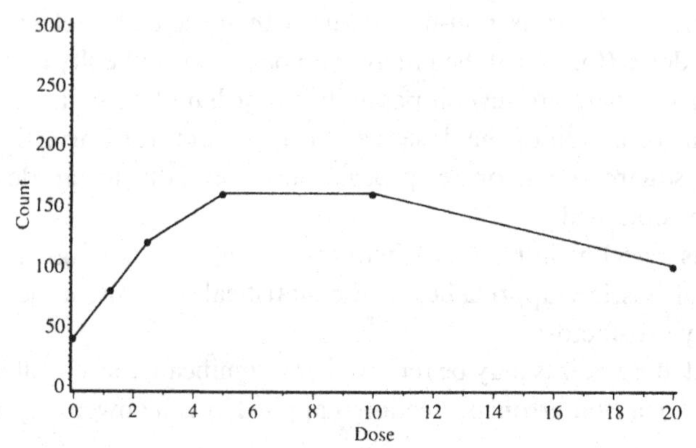

- extra parameters may be included in the model; the significance level may be affected if several parameters are simultaneously estimated from the data and the estimates are not statistically independent; the significance level may be affected if the model is inappropriate, for example if an exponential toxicity model is assumed but a resistant sub-population is present;
- a multiple-comparisons method (e.g. Dunnett's method) may be used; some power will be lost.

In extreme cases there may be no alternative to repeating the experiment with a range of non-toxic doses.

2.3.1.5 *Transformations*

There are three roles for transformations in the analysis for Ames test data: the stabilisation of variances, the normalisation of data, and the linearisation of the dose response curve. The commonest transformations used to stabilise variance are square root and logarithm. The square root transformation is appropriate when the variance of the data equals the mean or a constant times the mean. The logarithmic transformation is appropriate when the variance equals a constant times the mean squared. The negative binomial distribution has been proposed to describe the distribution of colony counts (Margolin *et al.*, 1981). The inverse hyperbolic sine tranformation is appropriate for that distribution for which the variance equals the mean plus a constant times the mean squared.

In general, a power transformation may be effective in stabilising variance (Snee & Irr, 1984). For data with subclasses across which the variance s^2 varies with the mean $\bar{x}$, $\log s^2$ may be regressed upon $\log \bar{x}$. Let the slope of the regression line be b. If the data are transformed as follows: $z = x^p$ where $p = 1 - b/2$, the subclass variances and means of the transformed data will often be found to be unrelated. If p equals zero, z should be calculated as $\log x$. One difficulty in using this method of choosing a power transformation with Ames test data is that the usual three-fold replication is insufficient to give a good estimate of s^2. Data from several similar experiments may be combined, pooling s^2 values for groups with similar means.

The second purpose of transforming the data is to make their distribution closer to a normal distribution. This is relevant for parametric tests where such a distribution is assumed. The transformation used to stabilise variance may also render the data more normal. Similar conclusions will often be drawn no matter which of the common transformations is used.

The third use of transformation is to make the dose-response curve more linear. Several transformations of the dose scale are commonly used including logarithms and reciprocals. With these transformations an arbitrary constant will need to be added to each dose value to avoid difficulty with the zero dose (control).

2.3.1.6 *Weighting*

For regression when subgroup variances are not homogeneous, a weighted analysis may be performed as an alternative to the use of a transformation. The weighted analysis has the advantage that the regression coefficient may be directly interpreted as the change in number of colonies per unit change in dose. In general, each data point should be weighted by the reciprocal of its true subgroup variance. In practice, the true value is not known and three data values per dose subgroup are not sufficient to give a good estimate of variance. Each datum point may be weighted by the reciprocal of its subgroup mean, though the use of any reasonable weighting scheme should lead to similar results.

2.3.1.7 *Repeated experiments*

The standard protocols for the Ames and similar tests specify a certain number of dose levels, appropriately spaced, and a certain number of replicates per dose level and control. In addition, it is often recommended that for an agent to be judged positive, a statistically positive result should be obtained in two separate experiments. If the two experiments yield conflicting results a third should be performed. Thus six overall outcomes are possible: + +, + − +, − + +, + − −, − + − and − −; the first three being taken to indicate a positive agent.

The recommendation has implications for the power and sensitivity of the test. Two experiments combined will be more powerful than one alone, and the third experiment implies an increase in scale and thus in power in borderline cases.

The recommendation also has implications for the significance level. If the usual 5% level is used for a single experiment, the probability of each of the six outcomes for a negative agent is: 0.0025, 0.002375, 0.002375, 0.045125, 0.045125, and 0.9025 respectively. Thus the probability of a false positive outcome overall is 0.00725, implying a significance level of 0.725%.

2.3.1.8 *The doubling rule*

It is frequently recommended that, for an agent to be judged positive, it should result in at least a doubling of the control frequency of revertant colonies. For a low control frequency this rule may correspond to statistical significance. Assuming a Poisson distribution and a control count of 25, a treatment count of 50 will be just significant at the 5% level. However, when the control frequency is larger the doubling rule may be seen to be very conservative. With a control count of 250, a treatment count of 320 (much less than 500) will be just significant. The doubling rule has also been discussed and criticised by Chu *et al.* (1981), Mitchell (1982) and Carnes *et al.* (1985).

2.3.1.9 *Additive and multiplicative effects*

Researchers are often urged to reduce the background frequency as much as possible. This advice depends on protocol and mutagenic effects being additive when they may sometimes be multiplicative. Consider the following data:

	Colony count	
	Control	Treatment
Experiment 1	30	60
Experiment 2	15	45
Experiment 3	15	30

In Experiment 1 the treatment count is twice the control count and, if the counts have a Poisson distribution, the treatment effect is statistically significant. Experiment 2 is a variant of experiment 1 in which the control count is reduced by 15. This is an example of what would happen if pre-existing mutants constituted half the control number in Experiment 1 and if they were eliminated in Experiment 2. The change in experimental procedure also reduces the treatment count by 15, an additive effect. The treatment count is now three times the control and is more highly significant than before. Hence the advice that background counts should be minimised.

Experiment 3 is another variant of Experiment 1 in which the control count is reduced to 15. However, the treatment count is now 30. This might arise if only half as many cells were exposed to the test agent in Experiment 3 compared to Experiment 1. Both treatment and control counts are halved, a multiplicative effect, and the difference is not statistically significant.

2.3.1.10 *Bonferroni correction*

When an experiment consists of a control group and a single treatment group, a significance level of 5% implies a 1 in 20 probability that the two group means will be judged to be different when there is no underlying difference. When an experiment consists of several pairs of treatment and control groups and the difference between each pair of means is tested at the 5% level, the probability of finding *at least one* significant difference due to chance is much more than 1 in 20. In fact, with five independent tests the probability is about 23%.

There are two approaches to arrive at an appropriate significance level for an experiment as a whole. The size of the difference to be judged significant may be adjusted, or the significance level of the individual comparisons may be adjusted (reduced) to take account of the multiple comparisons. Dunnett's method (Sections 2.3.2.1, 2.3.3.3) is a widely used example of the first approach and the Bonferroni correction is an example of the second.

The Bonferroni correction consists of dividing the required significance levels for an experiment by the number of independent comparisons in the experiment and using the resulting value as the significance level for each comparison. For example, if an experiment consists of ten pairs of treatment and control groups, and if a significance level of 5% for the experiment is desired, then each treatment group should be compared to its control using a significance level of $\frac{5}{10}$ or 0.5%.

The Bonferroni correction may still be used when the tests within an experiment are not independent. For example, when an experiment consists of a control group and five dose groups, each dose group may be compared to the control using a significance level of 1%. The overall significance level will be approximately 5% though less than 5%. The procedure is conservative and there will be a tendency for true weak effects to be judged non-significant. An improved procedure for use in such circumstances has been described by Simes (1986).

2.3.1.11 *Number of doses and replicates*

The usual experimental design of five dose levels and triplicate plates may be seen as a compromise among several ideals. Five dose levels plus control and three plates per level imply 18 plates per experiment and may be termed a 6×3 design. This number of plates may be arranged in four possible balanced designs, i.e. 2×9, 3×6, 6×3 and 9×2, each with different properties.

The 2×9 design consists of controls and one dose level with nine-fold replication. If the dose level can be chosen at the dose of maximum effect then the design is the most powerful possible for detecting a true difference between treatment and control counts. The number of error degrees of freedom is 16 allowing a good estimate of variance within experiment. The problems with this design are that, unless one or more preliminary experiments have been performed, one does not know the dose of maximum effect, and the design yields no information on the form of the dose-response. The 3 × 6 design (controls, two dose levels, six plates per level) has 15 error degrees of freedom but provides little information on the form of the dose response.

The usual 6 × 3 design has 12 error degrees of freedom. The five dose levels (plus control) give more detailed information on the form of the dose-response. If there is a marked decline in counts at one or two top doses, perhaps due to toxicity, these counts may be omitted and sufficient data remain to demonstrate a dose-related effect.

The 9 × 2 design (controls, eight dose levels, duplicate plates) is not a powerful design for detecting a weak dose effect since so many plates are assigned to intermediate doses. There are only nine degrees of freedom for estimating the within-experiment variance. However, if there is a dose-related effect the many doses will give much information on the form of the dose response.

There is one departure from a balanced design which is advantageous: the replication of the controls should be greater than that of the dose groups. With multiple comparison methods of analysis, the mean count for each dose is compared to the control mean and the statistical test will be more sensitive if the control mean is based on more plates. With regression methods, provided that the dose range is adequately covered, assigning extra plates to the controls is more efficient than assigning them to an intermediate dose. A nearly optimal allocation of plates is for the control to have twice as many plates as each dose. In some laboratories it is usual to begin and end each experiment with a control group.

2.3.2 Outline of published methods

Two basic approaches may be distinguished for determining statistical significance. One may compare control versus treatment values (individually or collectively) or one may seek a progressive dose-response. Either approach is amenable to parametric or non-parametric methods of analysis. For the parametric methods, it is assumed that the data come from a population with a particular distribution and certain parameters such as variance, mean and regression coefficient are to be estimated from the actual data. For non-parametric methods, simpler assumptions of uniformity are made.

Parametric methods usually make more assumptions than non-parametric methods. When the assumptions are correct the parametric methods are more powerful (i.e. are less likely to lead to false negative conclusions) than non-parametric ones. However, when the assumptions are not correct the significance level of parametric tests may be distorted and an inflated number of false positive conclusions may be reached. The non-parametric tests do have limitations. The methods based on ranking assume that control and treatment data come from distributions with the same shape. Some non-parametric methods may require more sample replications than parametric methods.

2.3.2.1 *Parametric multiple comparisons methods*

In an experiment with controls and one dose group the average colony count for the controls may be compared with that of the dose group using Student's *t*-test. With two or more dose groups, separate pairwise comparisons may be made as suggested by Weinstein & Lewinson (1978). These authors transformed the colony counts to square roots to stabilise variance and used the Bonferroni correction. With this method, if a particular dose group has more variable data the variance of the difference from control is increased and hence a larger difference from control is needed before statistical significance is achieved. Small increases from control may be significant and larger ones not.

If the estimates of variance for each group do not differ too much (perhaps after a transformation), the values may be pooled into a single estimate which can be used to calculate a Student's *t*-statistic for each comparison. Tables for testing such *t*-statistics and which cater for multiple comparisons have been published by Dunnett (1955, 1964) and a worked example of Dunnett's methods is presented in Section 2.3.3.3.

A multiple comparisons method which takes account of any progressive response across doses has been reported by Williams (1971, 1972), and a non-parametric version of this method has been reported (Shirley, 1977; Williams, 1986).

2.3.2.2 *Regression methods*

Several authors have used regression methods to analyse colony count data. The differences in method used depend on whether any downturn in the dose-response is excluded or modelled, and the choice of transformation or weighting scheme.

(a) *Linear regression*

The use of this method assumes that the dose-response relationship is a straight line and that variability about the line is constant for the controls and across the dose groups. The validity of the assumptions may be assessed from a graph of the data. If they are invalid, a transformation of the counts may stabilise variances and a transformation of dose may linearise the dose-response. The regression coefficient indicates the change in (transformed) count due to a unit change in (transformed) dose. Colony counts have been analysed untransformed by Venitt & Croften-Sleigh (1979). Counts were transformed to logarithms by Chu *et al.* (1981). Agnese *et al.* (1984) transformed both the counts and doses to logarithms. A worked example of linear regression is presented in Section 2.3.3.2.

In the presence of a downturn in the dose-response, several authors have advocated removing the dose points so that an essentially linear relationship remains (Venitt & Crofton-Sleigh, 1979; Chu *et al.* 1981; Bernstein *et al.* 1982; Moore & Felton, 1983). Different authors have used different procedures to decide which data to exclude. Moore & Felton (1983) take a pragmatic approach using weighted linear regression and rely on graphs of the data to indicate when to remove a dose level. Bernstein *et al.* (1982) provide a more formal approach validated by simulation results. Successive linear fits are followed by tests at a suggested 25% significance level to decide whether to remove the top dose. Mutagenicity and goodness of fit are tested, and a computer is needed for the calculation. The exclusion of data when there is an apparent downturn can lead to a distortion of the significance level and an increased frequency of false positives. Bernstein *et al.* (1982) propose a Bonferroni-type correction to overcome this problem.

(b) *Isotonic regression*

In linear regression it is assumed that a constant increase in (transformed) dose yields a constant increase in (transformed) colony count. In isotonic or order-preserving regression the constant increase is not necessary as the less restrictive assumption is made that an increase in dose yields either an increase or no change in colony count. The control should be included in this regression which involves fitting values to the observed counts such that they either increase or remain the same with each increase in dose. Barlow *et al.* (1972) describe the 'Pool the Adjacent Violators' algorithm for determining fitted values. They also describe a test of significance for the dose-response relationship.

(c) *Modelling*

An alternative approach for dealing with toxicity is to assume it has a particular mathematical form. Models have been proposed by several authors to account for the observed mutational yield when unicellular micro-organisms are exposed to a mutagenic agent. Based on single-hit kinetics, Eckardt & Haynes (1980) suggested a model composed of two exponential terms multiplied together. The first (positive) term is associated with mutagenicity and the second (negative) one with toxicity. Such modelling has been used by Stead *et al.* (1981), Margolin *et al.* (1981), Myers *et al.* (1981) and Imamura *et al.* (1983). These methods all require a computer for calculation and all depend on the validity of the underlying mathematical model.

Stead *et al.* (1981) assumed a Poisson distribution for the colony counts and estimated parameters of the model by maximum likelihood. Log-likelihood ratios were used to test the significance of the mutagenicity and toxicity estimates, the goodness of fit (adequacy) of the model and the assumption of the Poisson distribution.

Margolin *et al.* (1981) examined generalisation of the model and proposed the negative binomial distribution to represent sampling variation. Log-likelihood ratios were used to assess the goodness of fit and several variants of the model fitted some data almost equally well.

Myers *et al.* (1981) approximated the mutagenicity curve by a straight line, and sampling variation was assumed to depend on group mean through a power relationship. The parameters of the model were estimated by iteratively reweighted least squares. The method may be extended to reduce the weight given to the outlying points resulting in a robust analysis.

2.3.2.3 *Non-parametric methods*

The non-parametric methods proposed for analysing mutagenicity data include the Kolmogorov-Smirnov method and various ranking methods. Mitchell (1982) advocated the Kolmogorov-Smirnov two-sample test. The cumulative frequency of mutant colonies across control plates is compared to the cumulative frequency for each dose level. The test statistic is the maximum difference between the cumulative frequencies, a Bonferroni correction to the significance level being used when there are several dose levels. The method was illustrated with data from an experiment with eight control plates and four plates per dose level.

Boyd (1982) presented three ranking methods: aligned rank order test, Chako's test and Jonckheere's test, applied to data from the mouse lymphoma assay. It is clear that the methods are equally applicable to data from the microbial assays.

Wahrendorf *et al.* (1985) presented a ranking method for testing the significance of the dose-response. All the data from an experiment are ranked in ascending order. If a 'cutpoint' is taken between two dose levels the data are divided into two sets. The first cutpoint defines the sets: the controls and the treatment data. The second cutpoint defines the sets: the controls with the data for the lowest dose, and the rest of the treatment data; and so on. The sum of ranks may be calculated for the set of data below each possible cutpoint, and the test statistic is the total of all such sums. The significance of the test statistic is assessed using tables derived by Monte-Carlo simulation, or by approximation to the normal distribution. A worked example of this method is presented in Section 2.3.3.4.

Simpson & Margolin (1986) proposed a recursive ranking method. The method is specially designed to take account of a possible downturn in response at high doses. They assess the power of their method in comparison with other non-parametric methods by means of computer simulation. In the *absence* of a downturn, methods such as Jonckheere's method were found to be most powerful, followed by the method of Simpson & Margolin, and followed by certain non-parametric multiple comparison methods. However, in the *presence* of a downturn at high doses, the method of Simpson & Margolin was found to be most powerful, followed by the multiple comparisons methods, and followed by the Jonckheere and similar methods.

2.3.3 Worked examples

An example is presented below of the calculation of the χ^2 statistic and the analysis of a set of data using one method from each of the Sections 2.3.2.1, 2.3.2.2 and 2.3.2.3. In general, the examples are only developed to the point of testing whether or not the agent gives rise to a significant effect. The papers reviewed in Section 2.3.2 indicate the wider range of analyses that are possible. The data in the worked examples are from an experiment carried out by A.M. Camus and discussed briefly by Bartsch *et al.* (1980). The test agent was a pesticide (compound G24), and *Salmonella* strain TA100 and a rat-liver microsomal preparation were used. A graph of the data is shown in Fig. 2.2.

Fig. 2.2. Data from an experiment with a pesticide using strain TA100 with activation.

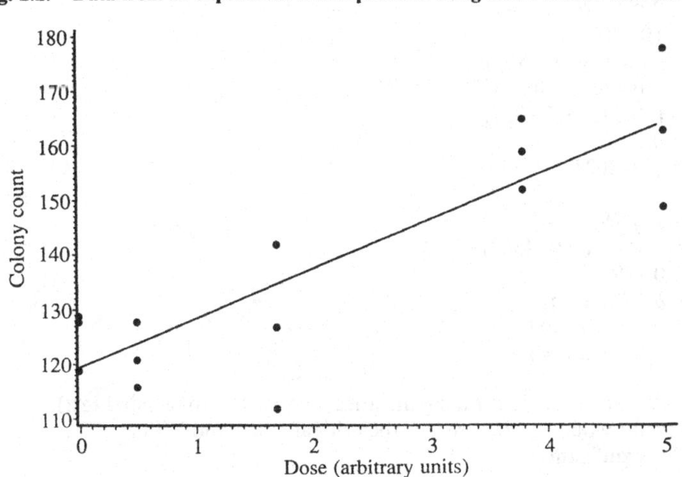

Box 2.3. *Worked example: linear regression*

Dose d	Colony count $x(=z^2)$	Transformed count z	Dose squared d^2	Cross product dz
0	129	11.3578	0	0
0	128	11.3137	0	0
0	119	10.9087	0	0
0.5	128	11.3137	0.25	5.6569
0.5	116	10.7703	0.25	5.3852
0.5	121	11.0000	0.25	5.5000
1.7	127	11.2694	2.89	19.1580
1.7	142	11.9164	2.89	20.2578
1.7	112	10.5830	2.89	17.9911
3.8	165	12.8452	14.44	48.8119
3.8	152	12.3288	14.44	46.8495
3.8	159	12.6095	14.44	47.9162
5.0	178	13.3417	25.00	66.7083
5.0	149	12.2066	25.00	61.0328
5.0	163	12.7671	25.00	63.8357
$\sum d=33$	$\sum z^2=2088$	$\sum z=176.5320$	$\sum d^2\,127.74$	$\sum dz=409.1034$

$$
\begin{aligned}
SCP &= \textstyle\sum dz - \sum d \sum z/n \\
&= 409.1034 - (33 \times 176.5320/15) \\
&= 409.1034 - 388.3705 \\
&= 20.7330 \\
SS_d &= \textstyle\sum d^2 - (\sum d)^2/n \\
&= 127.74 - (33)^2/15 \\
&= 127.74 - 72.60 \\
&= 55.14 \\
b &= SCP/SS_d = 0.3760 \\
SS_z &= \textstyle\sum z^2 - (\sum z)^2/n \\
&= 2088 - (176.5320)^2/15 \\
&= 2088 - 2077.5698 \\
&= 10.4302 \\
s^2 &= (SS_z - SCP^2/SS_d)/(n-2) \\
&= (10.4302 - 20.7330^2/55.14)/(15-2) \\
&= (10.4302 - 7.7957)/13 \\
&= 0.2027 \\
s &= \sqrt{(0.2027)} = 0.4502
\end{aligned}
$$

S.E. of

$$
\begin{aligned}
b &= s/\sqrt{SS_z} \\
&= 0.4502/\sqrt{(10.4302)} \\
&= 0.1394 \\
t &= b/(S.E. \text{ of } b) \\
&= 0.3760/0.1394 \\
&= 2.6973 \quad d.f. = n-2 = 13
\end{aligned}
$$

Critical value of t is 2.650 for a significance level of 1% (one-sided test). The observed t value, 2.697, *exceeds* the critical value and is thus statistically significant.

Box 2.3. (*cont.*)

$\bar{z} = a + b\bar{d}$
$\bar{z} = \sum z/n = 176.5320/15 = 11.7688$
$\bar{d} = \sum d/n = 33/15 = 2.2$

11.7688	$= a + (0.3760 \times 2.2)$
	$= a + 0.8272$
a	$= 11.7688 - 0.8272$
	$= 10.9416$
a^2	$= 119.7186$
$a+b$	$= 10.9416 + 0.3760 = 11.3176$
$(a+b)^2$	$= 128.0881$
$(a+b)^2 - a^2$	$= 128.0881 - 119.7186$
	$= 8.3695$

Treat and plate assays, with the treat step in liquid culture, are often performed with duplicate cultures and triplicate plates per culture. The variation within triplets gives an underestimate of sampling variation since between-culture variation is not included. The counts for the triplicate plates should be averaged, after which the duplicate averages may be analysed by the methods illustrated in Sections 2.3.3.2, 2.3.3.3 and 2.3.3.4.

2.3.3.1 *The χ^2 and* m *statistics*

The first step in the calculation of this statistic is the accumulation of the sums ($\sum x$) and the sums of squares ($\sum x^2$) of the colony counts for the controls and the dose groups, as shown in Box 2.1. The next step is the calculation of the corrected sum of squares (SS_X) for each group according to the formula $SS_X = \sum x^2 - (\sum x)^2/n$, where n is the number of counts in the group. The mean count ($\bar{x}$) is then calculated and $\bar{x} = \sum x/n$. Finally, the χ^2 value for each group is calculated as $SS_X/\bar{x}$ with number of degrees of freedom (*d.f.*) equal to $n-1$. The χ^2 values may be added together to give a single value for the experiment, with number of degrees of freedom equal to the sum of the *d.f.* values.

The m statistic may be calculated as χ^2 divided by its degrees of freedom. The m value in the example is 0.77, that is, the variability of the data is about 23% less than that expected from the Poisson distribution. On the basis of this value it would not be unreasonable to use a method of analysis which assumed that distribution. In fact, no such assumption is made in the following examples.

2.3.3.2 *Linear regression*

The first step with this method is to transform the data to stabilise variances (Box 2.3). The transformed variable in the example is z, the square root of the colony count x. The table also shows d, the dose and d^2, its square and dz, the product of the dose and the transformed count. The next step is to calculate the corrected sum of cross products, SCP, between dose and transformed count. The value of SCP is 20.7330. Next one calculates SS_d, the corrected sum of squares of doses. The value of SS_d is 55.14.

The coefficient of regression of transformed count on dose, b, is the ratio of SCP to SS_d. The value of the regression coefficient is 0.3760, i.e. an increase of one unit of dose is associated with an increase of 0.3760 units on the transformed count scale. The corresponding increase on the untransformed scale is considered below.

To test the significance of b, the t statistic must be calculated. The first step is to calculate SS_z, the corrected sum of squares of the transformed counts. The variance s^2, of the transformed counts about the regression line, and the standard deviation, s, may then be calculated.

The t statistic equals b multiplied by the square root of SS_z and divided by s. The value of t in the example is 2.6978 and its number of degrees of freedom is two less than the number data points, i.e. 13. A one-tailed test of t is appropriate since the test compound is expected, if anything, to give a positive dose response. Critical values of t are usually tabulated for a two-tailed test. Consulting such a table, the critical value of t with 13 degrees of freedom and significance level of 2% is seen to be 2.650, thus the critical t value for a one-tailed test and a significance level of 1% is 2.650. The observed t value is 2.698 and it exceeds the critical value, thus one concludes that the observed value is statistically significant at the 1% level.

The regression coefficient, b, is an indirect measure of the dose response. A more direct measure may be calculated as follows. The average transformed count, $\bar{z}$, equals the intercept, a, of the regression line on the z axis, plus b times the average dose $\bar{d}$. Hence a may be calculated; its value in the example is 10.9416. The difference between the square of $a+b$ and the square of a is approximately equal to the slope of the dose response at low doses on the untransformed count scale. The difference in the example was 8.3696, that is, an increase in dose of 1 unit is equivalent to about eight more revertant colonies.

2.3.3.3 *Dunnett's method*

The first step is to tabulate z, the square root of each colony count, x (Box 2.4). The square root transformation is applied to stabilise variances. The next step is to calculate the sum of the z values, $\sum z$, and their squares, $\sum z^2$, for the controls and each of the doses. The corrected sum of squares of transformed counts, SS_z, may then be calculated for each case.

Box 2.4. *Worked example: Dunnett's method*

Dose d	Colony count $x(=z^2)$	Transformed count z	$\sum z$	$\sum z^2$	Corrected sum of squares SS_z	$d.f.$	Average transformed count $\bar{z}$	t
0	129	11.3578						
0	128	11.3137	33.5802	376.	0.1234	2	11.1934	—
0	119	10.9087						
0.5	128	11.3137						
0.5	116	10.7703	33.0840	365.	0.1496	2	11.0280	−0.46
0.5	121	11.0000						
1.7	127	11.2694						
1.7	142	11.9164	33.7688	381.	0.8894	2	11.2563	0.017
1.7	112	10.5830						
3.8	165	12.8452						
3.8	152	12.3288	37.7835	476.	0.1357	2	12.5945	3.89[a]
3.8	159	12.6095						
5.0	178	13.3417						
5.0	149	12.2066	38.3154	490.	0.6434	2	12.7718	4.39[a]
5.0	163	12.7671						
					1.9415	10		

$SS_z = \sum z^2 - (\sum z)^2/r$ (r, replication factor)

$EMS = SS_z/d.f.$
$= 1.9415/10$
$= 0.1942$

$t = (\bar{z}_d - \bar{z}_c)/\sqrt{(2.EMS/r)}$
$= (12.7718 - 11.1934)/\sqrt{(2 \times 0.1942/3)}$
$= 1.5784/0.3598$
$= 4.387$

Note:
[a] Critical value t is 3.45 for a significance level of 1% (one-sided test). Two observed t values, 3.89 and 4.39, *exceed* the critical values and are thus statistically significant.

Each sum of squares has a number of degrees of freedom equal to the number of replicates minus 1.

The error mean square, *EMS*, may be calculated as the sum of the SS_z values divided by the sum of their degrees of freedom. In the example, *EMS* has the value 0.1939. The average transformed count, $\bar{z}$, for the controls and each level may be calculated and the difference between the control mean and each dose mean may be calculated. Each such difference divided by its standard error is a *t*-statistic. The standard error of the difference is the square root of twice the error mean square divided by the replication factor, r.

Critical values of *t* when several treatment means are compared to a control and, where the test is one sided, may be found in Tables 1a and 1b of Dunnett (1955). For four treatment means and control, 10 degrees of freedom, and a significance level of 1%, the critical value of *t* is 3.45. The largest observed *t* value is 4.39 and it exceeds the critical value, so one concludes that a statistically significant effect due to dose has been observed. Another *t* value is also significant, that for the dose of 3.8. Thus one may conclude that in this experiment the lowest effective dose was 3.8 units.

2.3.3.4 *Wahrendorf's method*

The first step with this non-parametric method is to rank the colony counts in ascending order and the ranks, *R*, are shown in Box 2.5. The sums of ranks up to each cutpoint are calculated and these sums form the basis of the test statistic. The ranks for the plates in each dose group are added ($\sum R$). The sum, L_1, to the first cutpoint is then the $\sum R$ for the controls, 17.5. The sum, L_2, to the second cutpoint is the sum of ranks for the controls and the lowest dose group, $17.5+12.5=30$, and so on. No L_j value should be calculated for the top dose. A check on the calculations can be made by calculating the sum of ranks for all dose groups and this should equal $\frac{1}{2}n(n+1)$ where n is the number of colony counts. Each L_j is a Wilcoxon Rank Sum test statistic for comparing the dose groups up to the jth cutpoint with those beyond it.

Wahrendorf's test statistic is the sum of the L_js. In the example its value is 174.5. Critical values of *L* for various replication factors and numbers of dose levels may be found in Table 2 of Wahrendorf *et al.* (1985). In an experiment with four dose levels and three-fold replication for the controls and each dose level, the critical value of *L* is 184 with a significance level of 1%. Since the calculated *L*, 174.5, is *smaller* than the critical value, one may conclude that the test statistic is significant; one may conclude that there is a significant dose-response.

Box 2.5. *Worked example: Wahrendorf's method*

Dose d	Colony counts x	Ranks R	Sum of ranks $\sum R$	Sum to cutpoint L_j
0	129	8		
0	128	6.5	17.5	17.5
0	119	3		
0.5	128	6.5		
0.5	116	2	12.5	17.5 + 12.5 = 30
0.5	121	4		
1.7	127	5		
1.7	142	9	15	30 + 15 = 45
1.7	112	1		
3.8	165	14		
3.8	152	11	37	45 + 37 = 82
3.8	159	12		
5.0	178	15		
5.0	149	10	38	—
5.0	163	13		
				$L = 174.5$

Note:
Critical value of L is 184 for a significance level of 1% (one-sided test). The observed L value is *less* than the critical value and is thus statistically significant.

2.3.4 Computer simulation

2.3.4.1 *Introduction*

The power of a statistical test, defined in Section 2.3.1.2, gives a criterion for comparing experimental protocols and methods of statistical analysis. Statistical theory may be used to derive the power of a test at a particular significance level on the basis of a series of assumptions, particularly concerning the nature of sampling variation. However, in many practical situations the derivation of power from statistical theory is not available. In these situations, when there are complications in the protocol or calculations, or when simple assumptions about variation are invalid, computer simulation may be used. Many sets of simulated negative data may be generated and analysed to determine the actual significance level of a proposed test for a particular nominal level. Many sets of simulated positive data (often weakly positive) may be analysed to determine the power of a proposed test for a particular significance level. Such power estimates may be used to compare protocols and tests, even quite complicated ones.

Computer simulation has a number of limitations. Many variants of the experimental protocols and many statistical tests have been proposed and many hypotheses have been put forward about the nature of sampling variation. There are simply too many combinations for the simulator to consider them all, and combinations of practical importance may be among those he has chosen not to consider. Furthermore, some statistical tests may involve so much calculation that their application to many sets of simulated data may be computationally impracticable.

Nevertheless, simulation has been used to compare some of the methods of analysis and transformations considered in previous sections. Various control colony counts and levels of variability and a progressive dose response were included. However, a decline in the dose-response curve at high doses due to toxicity was not included.

2.3.4.2 *Methods*

Computer simulation was undertaken to compare three methods of analysis: linear regression, Dunnett's test and Wahrendorf's test. Three transformations were used: inverse hyperbolic sine, logarithm and square root. The simulated data were also analysed untransformed.

Three levels of background control counts were included: 4, 15 and 60 colonies per plate. The negative binomial distribution was chosen for the data. The variance, V, and mean, m, of this distribution are related as follows:

$$V = m + cm^2$$

When the constant c equals zero, the negative binomial distribution reduces to the Poisson distribution. Values of c were chosen such that V would exceed m in the controls by 20, 50 or 100% for each value of m. For example, with m equal to 15 and a percentage increase in variance of 50%, a c value of 1/30 was used. The same c value was used for control and treatment data in each simulated experiment.

The simulated experiments comprised control plates and five dose levels, with doses of 0, 1, 2, 4, 8 and 16 units respectively. Three-fold replication was assumed. There was a dose-related increase in mean colony count of 1.0, 1.2, 1.5, 2.0 and 2.5 times the control count.

For each combination of control mean and level of variability, 1500 sets of simulated data were generated. Using the various combinations of method of analysis and data transformation, each set of simulated data was analysed nine times. A one-sided test of significance at the 5% level was made in the direction of a positive dose response.

2.3.4.3 *Results*

An initial simulation of negative data was used to check the significance level of the methods. 1500 sets of data were generated and analysed using a significance level of 5%. All the methods of analysis yielded about 5% significant results, indicating that the actual significance level of each method closely corresponds to the nominal one. The results of the analysis of simulated positive data are summarised in Box 2.6. Each cell in the table contains the percentage of the 1500 sets of simulated data that showed a statistically significant effect of dose.

For the highest control level, i.e. 60 colonies per plate, 100% of sets of data gave a significant result irrespective of the method of analysis or transformation used. When the control level was 15 colonies per plate, the proportion of significant results ranged from 80 to 100%. For the lowest control level, four colonies per plate, the proportion ranged from 27 to 88%. In summary, when the dose effect was a multiple of the control level the greatest frequency of significant results was found with the highest control level (cf. Section 2.3.1.9).

The more variable sets of data gave a lower proportion of significant results. For example, with a control level of four colonies and an excess variance of 20, 50 or 100%, the proportion of significant results was 86, 75 or 59% respectively (linear regression, inverse hyperbolic sine transformation).

The various methods of analysis and transformations gave similar proportions of significant results, except with Dunnett's method with a low control level when the proportion of significant results was considerably below that given by other methods.

2.3.4.4 *Discussion*

The simulations provide the means to choose among transformations and methods of analysis. Three transformations were considered: inverse hyperbolic sine, logarithm, and square root. Raw, untransformed data were also analysed. The simulated data were generated so as to have a negative binomial distribution for which the inverse hyperbolic sine tranformation is appropriate, and this transformation is the basis for comparison. This transformation may be used in two variants, with the parameter c fixed in advance, or with c estimated from the data.

It may be seen from Box 2.6 that the smallest number of significant effects

Box 2.6. *Percentage of sets of simulated data giving significant results*

Control	PIV / Mean	20			50			100		
		4	15	60	4	15	60	4	15	60
Regression	IHS	86	100	100	75	100	100	59	99	100
	L	86	100	100	75	100	100	61	98	100
	SR	87	100	100	77	100	100	63	99	100
	—	88	100	100	79	100	100	66	99	100
Dunnett	IHS	50	98	100	40	94	100	27	82	100
	L	46	97	100	40	92	100	35	80	100
	SR	51	99	100	44	95	100	32	84	100
	—	59	99	100	50	95	100	36	85	100
Wahrendorf		85	100	100	76	100	100	60	98	100

Notes:
PIV Percentage increase in control variance.
IHS inverse hyperbolic sine.
L logarithm.
SR square root.
— no transformation.

was usually obtained with the log transformation, followed by the inverse hyperbolic sine (with fixed c) and square root transformations. The largest number of significant effects was found with untransformed data. If the data have a negative binomial distribution then the log transformation will tend to over-correct the pattern in the variance while the square root transformation will tend to undercorrect it. The inverse hyperbolic sine with estimated c and power transformations (results not shown) gave a similar number of significant effects to the inverse hyperbolic sine with fixed c. In general, all the transformations (or no transformation) yielded similar results.

Three methods of analysis were compared: linear regression, Dunnett's test and Wahrendorf's test. The analyses of simulated data give an indication of the relative power of the methods: the more significant results the more powerful the method. It may be seen from Box 2.6 that linear regression and Wahrendorf's non-parametric method gave the greatest number of significant results. Dunnett's method, however, was noticeably less powerful, especially when colony counts were small and variability large. One limitation of this comparison of tests should be stressed. Toxicity, leading to a decline in colony counts at high doses, was not considered. Such toxicity would be expected to reduce the power of the regression method and Wahrendorf's method but not reduce the power of Dunnett's method to the same extent.

2.4. CONCLUSIONS

(a) As many microbial cells as is practical should be treated with the test agent.

(b) Although a period of growth following treatment is required for induced mutations to be expressed, this period should not be longer than necessary. A test for induced mutation among a small number of bacteria against a background of spontaneous mutation among a large number of bacteria will be needlessly insensitive.

(c) With the Plate Incorporation protocol and a stable agent, although reducing the number of bacteria plated may, in exceptional circumstances, give an actual increase in sensitivity by eliminating pre-existing mutants without reducing the number of bacteria available to undergo induced mutation, a negative result will not then be convincing evidence of non-mutagenicity.

(d) It may not be apparent from the plate counts that an inadequate number of bacteria has been treated. Thus it is important that the results presented make clear that sufficient cells were treated.

(e) Protocols must be designed to limit the number of pre-existing mutants in starting cultures. Where it is apparent from assay controls that starting cultures contained an excessive, atypical number of mutants (based on historical data), assays should be repeated. Indeed, the validity of the test should be confirmed by checking that the positive and negative control values are consistent in both mean and variance with accumulated historic values.

(f) Standard Plate Incorporation or Treat and Plate protocols do not measure spontaneous mutation rate but look for an increase over a spontaneous mutant frequency resulting from mutations arising before, during and after the period of treatment. Hence, although plate counts can be used for tests of significance, they can not normally be used to compare the induced mutation rate to the spontaneous mutation rate.

(g) Unless it is obvious that the test agent has had no effect, the data should be graphed to give a visual impression of the form of any dose-response and the pattern of variability.

(h) Methods of analysis which assume the Poisson distribution for testing significance are acceptable only when the *m* value for the experiment ($\chi^2/d.f.$) is less than or equal to one.

(i) When *m* is greater than one, a method in which the significance test is based on the observed variance of the data, or a non-parametric method is recommended.

(j) Three methods of analysis: linear regression, Dunnett's method and Wahrendorf's method, have been considered in detail and can be recommended. Each has its strengths and weaknesses and other methods are not excluded.

(k) Linear regression assumes that variance across doses is constant and that the dose response is linear. If the variance is not approximately constant then a transformation may be applied or a weighted analysis carried out. If the dose-response tends to a plateau then the dose scale may be transformed. If counts decline markedly at high doses then linear regression is inappropriate.

(l) Dunnett's method, perhaps with a transformation, is recommended when counts decline markedly at one or two high doses. However, when the dose-response shows no such decline other methods may be more powerful.

(m) Wahrendorf's non-parametric method avoids the complications of transformations of weighting and is about as powerful as any other method. However, it is inappropriate when the response declines much at high doses.

(n) The use of double controls, e.g. six plates rather than three, will lead to a more powerful test.

(o) Experiments must be repeated. When toxicity is observed in the first experiment, the toxic range of dose should be avoided in the second.

(p) When the two experiments lead to conflicting conclusions, a third 'decider' experiment should be performed.

2.5 REFERENCES

Agnese, G., D. Risso & S. De Flora (1984). Statistical evaluation of inter- and intra-laboratory variations of the Ames test, as related to the genetic stability of Salmonella tester strains. *Mutation Research*, **130**, 27–44.

Ames, B.N., J. McCann & E. Yamasaki (1975). Methods for detecting carcinogens and mutagens with the Salmonella/mammalian-microsome mutagenicity test. *Mutation Research*, **31**, 347–64.

Barlow, R.E., D.J. Batholomew, J.M. Bremner & H.D. Brunk (1972). *Statistical Inference under Order Restrictions*. Wiley, New York, pp.13–16.

Bartsch, H., C. Malaveille, A.-M. Camus, G. Martel-Planche, G. Brun, A. Hautefeuille, N. Sabadie, A. Barbin, T. Kuroki, D. Drevon, C. Piccoli & R. Montesano (1980). Validation and comparative studies on 180 chemicals with *S. typhimurium* strains and V79 chinese hamster cells in the presence of various metabolising systems. *Mutation Research*, **76**, 1–50.

Bernstein, L., J. Kaldor, J. McCann & M.C. Pike (1982). An empirical approach to the statistical analysis of mutagenesis data from the Salmonella test. *Mutation Research*, **97**, 267–81.

Boyd, N.M. (1982). Examples to testing against ordered alternatives in the analysis of mutagenicity data. *Mutation Research*, **97**, 147–53.

Campbell, R.C. (1974). Some non-normal distributions, in: *Statistics for Biologists*. 2nd edn, Cambridge University Press, pp. 293–342.

Carnes, B.A., S.S. Dornfeld & M.J. Peak (1985). A quantitative comparison of a percentile rule with a 2-fold rule for assessing mutagenicity in the Ames assay. *Mutation Research*, **147**, 15–21.

Chu, K.C., K.M. Patel, A.H. Lin, R.E. Tarone, M.S. Linharat & V.C. Dunkel (1981). Evaluating statistical analyses and reproducibility of microbial mutagenicity assays. *Mutation Research*, **85**, 119–32.

Dunnett, C.W. (1955). A multiple comparison procedure for comparing several treatments with a control. *Journal of the American Statistical Association*, **50**, 1096–121.

Dunnett, C.W. (1964). New tables for multiple comparisons with a control. *Biometrics*, **20**, 482–91.

Eckardt, F. & R.H. Haynes (1980). Quantitative measures of mutagenicity and mutability based on mutant yield data. *Mutation Research*, **74**, 439–58.

Green, M.H.L. (1981). The effect of spontaneous mutation on the sensitivity of the Ames test, in: A. Kappas (Ed), *Progress in Mutation Research*, Elsevier. North-Holland Biomedical Press, **2**, 159–65.

Green, M.H.L. & W.J. Muriel, (1976). Mutagen testing using TRP+ reversion in *Escherichia coli*. *Mutation Research*, **38**, 3–32.

Jackson, B.A. & R. Pertel (1986). Regulatory implications of Ames' mutagenicity assay using *Salmonella typhimurium*. *Fundamental and Applied Toxicology*, **7**, 1–16.

Imamura, A., Y. Kurumi, D. Danzuka, N. Kodama, R. Kawachi & M. Nagao

(1983). Classification of compounds by cluster analysis of Ames test data. *Gann*, **74**, 196–204.

Kier, L.E., D.J. Brusick, A.E. Auletta, E.S. Von Halle, M.M. Brown, V.F. Simmon, V. Dunkel, J. McCann, K. Mortelmans, M. Prival, T.K. Rao & V. Ray (1986). The *Salmonella typhimurium*/mammalian microsomal assay. A report of the US Environmental Protection Agency Gene-Tox Program. *Mutation Research*, **168**, 69–240.

Luria, S.E. & M. Delbrück, (1943). Mutations of bacteria from virus sensitivity to virus resistance. *Genetics*, **28**, 491–511.

McGregor, D., R.D. Prentice, M. McConville, Y.J. Lee & W.J. Caspary (1984). Reduced mutant yield at high doses in the *Salmonella*/activation assay; The cause is not always toxicity. *Environmental Mutagenesis*, **6**, 545–57.

Margolin, B.H. (1985). Statistical studies in genetic toxicology: A perspective from the US National Toxicology Program. *Environmental Health Perspectives*, **63**, 187–94.

Margolin, B.H., N. Kaplan and E. Zeiger (1981). Statistical analysis of the Ames *Salmonella*/microsome test. *Proceedings of the National Academy of Sciences USA*, **78**, 3779–83.

Margolin, B.H., K.J. Risko, M.D. Shelby & E. Zeiger (1984). Sources of variability in Ames *Salmonella typhimurium* tester strains; Analysis of the International Collaborative Study on 'Genetic Drift'. *Mutation Research*, **130**, 11–25.

Maron, D.M. & B.N. Ames (1984). Revised methods for the Salmonella mutagenicity test in: B.J. Kilbey, M. Legator, W. Nichols, C. Ramel (Eds). *Handbook of Mutagenicity Test Procedures*, 2nd edn, Elsevier Sci. Pub. BV Amsterdam, 93–140.

Mitchell, I. de G., P.A. Dixon, P.J. Gilbert & D.J. White (1980). Mutagenicity of antibiotics in microbial assays: problems of evaluation. *Mutation Research*, **79**, 91–105.

Mitchell, I. de G. (1982). Establishment of stochastic significance in Ames tests and its relevance to genetical significance. *Mutation Research*, **104**, 25–8.

Mitchell, I. de G. (1987). The interaction of statistical significance, biology of dose-response and test design in the assessment of genotoxicity data. *Mutagenesis*, **2**, 141–45.

Moore, D.M. & J.S. Felton (1983). A microcomputer program for analysing Ames test data. *Mutation Research*, **119**, 95–102.

Myers, L.E., N.H. Sexton, L.I. Southerland & T.J. Wolff (1981). Regression analysis of Ames test data. *Environmental Mutagenesis*, **3**, 575–86.

Parry, J., T. Brooks, I. Mitchell & P. Wilcox (1984). Genotoxicity studies using yeast cultures, in: *UKEMS Sub-committee on Guidelines for Mutagenicity Testing*, Report, Part II, Supplementary Tests; Mutagens in Food; Mutagens in Body Fluids and Excreta; Nitrosation Products, Ed. B.J. Dean, United Kingdom Environmental Mutagen Society, Swansea, 27–61.

Selby, P.B. & Olson, W.H. (1981). Methods and criteria for deciding whether specific-locus mutation-rate data in mice indicate a positive, negative, or inconclusive result. *Mutation Research*, **83**, 403–18.

Shirley, E. (1977). A non-parametric equivalent of Williams' test for contrasting increasing dose levels of a treatment. *Biometrics*, **33**, 386–9.

Simes, R.J. (1986). An improved Bonferroni procedure for multiple tests of significance. *Biometrika*, **73**, 751–4.

Simpson, D.G. & B.H. Margolin (1986). Recursive nonparametric testing for dose-response relationships subject to downturns at high doses. *Biometrika*, **73**. 589–96.

Snee, R.D. & J.D. Irr (1984). A procedure for the statistical evaluation of the Ames salmonella assay results. *Mutation Research*, **128**, 115–25.

Stead, A.G., V. Hasselblad, J.P. Creason & L. Caxton (1981). Modeling the Ames test. *Mutation Research*, **85**, 13–27.

Steel, R.G.D. & J.H. Torrie (1960). *Principles and Procedures of Statistics*. McGraw-Hill, New York.

Thompson, E.D. & P.J. Melampy (1981). An examination of the quantitative suspension assay for mutagenesis with strains of *Salmonella typhimurium*. *Environmental Mutagenesis*, **3**, 453–65.

Venitt, S. & C. Crofton-Sleigh (1979). Bacterial mutagenicity tests of phenazine methosulphate and three tetrazolium salts. *Mutation Research*, **68**, 107–16.

Venitt S., R. Forster & E. Longstaff (1983). Bacteria Mutation Assays, in: *UKEMS Sub-committee on Guidelines for Mutagenicity Testing*, Report, Part I, Basic Test Battery, B.J. Dean (Ed), United Kingdom Environmental Mutagen Society, Swansea, 5–40.

Wahrendorf, J., G.A.T. Mahon & M. Schumacher (1985). A non-parametric approach to the statistical analysis of mutagenicity data. *Mutation Research*, **147**, 5–13.

Weinstein, D. & T.M. Lewinson (1978). A statistical treatment of the Ames mutagenicity assay. *Mutation Research*, **51**, 433–4.

Williams, D.A. (1971). A test for difference between treatment means when several dose levels are compared with a zero dose control. *Biometrics*, **27**, 103–17.

Williams, D.A. (1972). The comparison of several dose levels with a zero dose control. *Biometrics*, **28**, 519–31.

Williams, D.A. (1986). A note on Shirley's nonparametric test for comparing several dose levels with a zero-dose control. *Biometrics*, **42**, 183–6.

Yahagi, T., M. Nagao, Y. Seino, T. Matsushima, T. Sugimura & M. Okada, (1977). Mutagenicities of N-nitrosamines on Salmonella. *Mutation Research*, **48**. 121–30.

3

Mammalian cell gene mutation assays based upon colony formation

C.F. ARLETT (Group Leader)
D.M. SMITH G.M. CLARKE
M.H.L. GREEN J. COLE
D.B. McGREGOR J.C. ASQUITH

3.1 INTRODUCTION

The mutation assays to be considered in this section depend upon the ability of mutant cells to form clones in the presence of toxic selective agents, and data are generated in the form of colony counts. Two classes of cells are used in these assays: (a) those which grow attached to glass or tissue culture grade plastic surfaces, these are represented by V79 and CHO Chinese hamster cells and (b) those which grow in stirred or suspension cultures, represented by L5178Y mouse lymphoma cells, from which colonies are established in semi-solid agar. A range of selective agents is available (Cole & Arlett, 1984), but many are specific to a particular cell type and only three are in widespread use. These are 6-thioguanine (6-TG), ouabain (OUA) and trifluorothymidine (TFT). Details of their mode of action, stringency of selection and routine handling have been discussed by Cole *et al.* (1983). A vital requirement is that the concentration of the selective agent should be high enough to kill all non-mutant cells.

The basic experiment (see flow diagram in Fig. 3.1) requires the provision of a large bulk culture of cells to be divided into control and treated sub-populations. The bulk population is maintained in such a way that a defined frequency of spontaneous mutants may be anticipated at the time of treatment. This 'historical' spontaneous or negative control mutant frequency is a useful guide to stability in the system. It also provides guidance as to the size of the populations to be treated and sampled.

At the termination of treatment, which usually takes three hours, cells are washed and cultured in fresh growth medium. An assessment of the toxic effects of any treatment is essential to determine if the dose levels chosen are appropriate. If the toxic effect is to be evaluated by measurements of cloning efficiency, samples of cells are plated in the non-selective medium. Up to 500 cells per dish may need to be plated from treated cultures where the cloning efficiency (= ratio of colonies formed/cells plated) is low. Two to six dishes are set up for each of these determinations of survival. An alternative method for estimating any toxic effect of a treatment is to undertake population counts during growth following treatment (Clive *et al.*, 1979). In mammalian mutation assays it is usual to recommend that no more than 90% kill is inflicted by the highest level of treatment.

Fig. 3.1. Flow diagram for a typical mutation assay for 6-thioguanine and ouabain resistance with L5178Y cells using 20 ml cultures in duplicate.

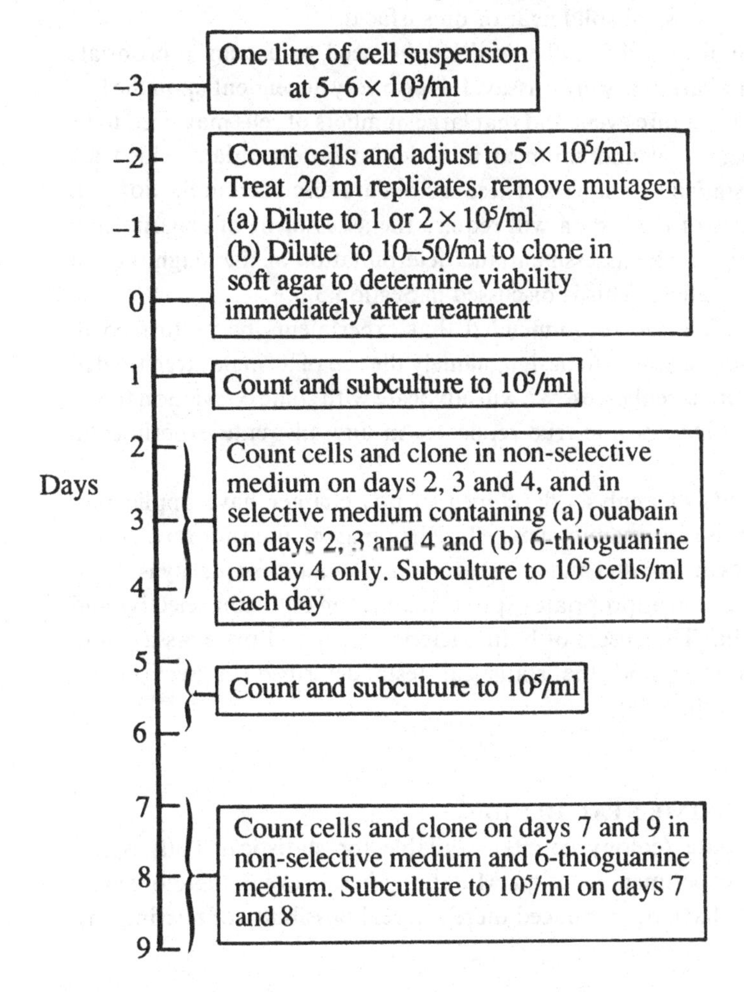

Time for expression of induced mutants is normally several days and can vary between selective agent, mutagen and cell type, but is usually determined by previous experience. Thus subculture of the control and treated populations is normally required. It is necessary here that cell numbers are recorded and that populations of *appropriate size are maintained.* The additional experimental error introduced through subculturing has important consequences for experimental design and is discussed in Section 3.2.3. below. When newly induced mutants are fully expressed, cells are plated at low (5–50 cells/ml) density in non-selective medium to determine cloning efficiency (viability) and at high (1×10^4–10×10^4 cells/ml) density in selective medium to determine mutant frequency. These plates are then incubated for 7–14 days, depending upon the growth rate of the cells, and scored for colony number. The criteria set for a colony to be included in the count depend upon colony size whether cells are grown in semi-solid agar or on surfaces.

Experimental design is influenced by the need to use an appropriate estimate of the variability in an assay. It is also very dependent upon the fact that mutation is a rare event and that large numbers of cells may need to be treated, subcultured and sampled in order to produce data which are suitable for statistical analysis. These factors are discussed in Section 3.2.

The analysis of these data will require the assessment of experimental quality, as well as the assessment and determination of the magnitude of any mutagenic effect. This is discussed in Section 3.3.

In 1983 Cole *et al.* recommended that experiments be performed to resolve a major dilemma in design, namely the use of a single large treated population versus replicates. We will advocate with some conviction that it is essential to incorporate true replicates in any adequate experimental design.

The methods of analysis developed in this chapter have application beyond their use for mammalian cells. They may be used for any system where cultures are treated and an assessment of mutant frequency is based on colony counts at appropriate expression times on both non-selective and selective media. Thus users of both bacterial 'treat and plate' assays and yeast assays may find the statistical tests described in Section 3.4 appropriate to their data.

3.2 EXPERIMENTAL DESIGN

The data (colony counts) available for statistical analysis are influenced by experimental design. Mutation is a rare event, therefore even when mutants have been induced there is a real possibility of missing any effect by either treating or sampling too few cells. A well-designed experiment takes this into account.

Another important point is to recognise the various sources of variability ('error') in the data and to obtain measures of all these for proper use in an analysis. Experiments with inadequate replication, or with no genuine assessment of 'between replicate cultures' variation, do not yield suitable data.

3.2.1 The number of cells to be treated and subcultured

The measurable spontaneous mutant frequency varies between cell types and selective systems. For each system steps must be taken to avoid the build-up of pre-existing spontaneous mutants. For example, acceptable spontaneous mutant frequencies for L5178Y mouse lymphoma cells are: 1×10^{-6} for OUA, 5×10^{-6} for 6-TG and 2×10^{-5}–10×10^{-5} for TFT. For Chinese hamster cells these frequencies become 4×10^{-7} for OUA and 1×10^{-6}–10×10^{-6} for 6-TG. These frequencies suggest that very large populations of cells are required in order to detect mutants. The rule often proposed is that each population should contain sufficient viable cells so that at least ten spontaneous mutants will be present. If populations containing at least ten spontaneous mutants cannot be maintained, it is doubtful if the experiment will be worth performing. Indeed Furth *et al.* (1981) propose that sufficient cells be treated so that at least 100 mutants should *survive* treatment. Although this latter recommendation is desirable, it is only practical for certain systems. For instance, if a test treatment reduced viability to 10%, in the case of OUA for L5178Y cells, 10^9 cells would have to be treated. The practical consequence of these recommendations is that there is a preference for suspension cultures as the test system since it is generally easier to obtain sufficient numbers of cells than with cultures which grow attached to a surface.

3.2.2 The number of cells to be sampled for mutants

At first sight, if a single plating of 10^6 cells in the presence of the selective agent reveals no mutants then it might be concluded that no mutations have occurred. However, if the real frequency of mutants is small, there is an appreciable probability of a zero count. Suppose that frequency of mutants cells is 1 per 10^6: then in 10^6 cells there is a probability e^{-MF} of finding zero mutants (assuming a Poisson distribution). Zero counts may occur in more than one plate, and even if n plates are used a complete set of 0's may still arise. The probability of this happening is e^{-nMF}, and is shown in Table 3.1 for a range of values $n=2$ to $n=8$ and survival at 100%, 50% and 10%. Plating 10^6 cells when there is only 10%

survival at the end of the expression period is equivalent to having only 10^5 cells with 100% viability.

It is clear that, when the mutant frequency is less than 0.5 mutants in 10^6, and survival is 10%, plating 10^6 cells even as many as eight times leaves an unacceptably high probability (0.670) of finding no mutants on any plate. If we require that the probability of finding all zero counts shall be less than 0.05, the minimum number of cells that should be plated for each of n plates ($n=2$ to $n=8$) is shown in Table 3.2.

The actual number of cells which can be challenged by a selective agent in a single plate differs between attached versus suspension cultures and is influenced by a number of factors including metabolic co-operation, whereby toxic metabolites are transferred from one cell to another via gap junctions (see, for instance, van Zeeland *et al.*, 1972). There is also a less clearly defined 'cell density' effect such that some selective agents cause an immediate cessation of cell division, while, for others, a round of replication may be required for toxicity to be expressed. Toxic effects may also be modified by the metabolic products of dying cells. More cells may be plated in an individual dish for suspension cultures because a third dimension is provided by the soft agar and because metabolic co-operation does not apply. In practice, for 6-TG up to 5×10^4 cells per ml, and for OUA up to 2×10^5 cells per ml can be used for L5178Y wild-type cells (Cole & Arlett, 1984). For TFT selection, up to 3×10^4 cells per ml are used for L5178Y TK$^+$/$-$ cells (Moore-Brown *et al.*, 1981). This means that the maximum number of cells which can be plated in 5 cm diameter petri dishes in 10 ml medium for L5178Y wild-type cells is 5×10^5 for 6-TG and is 2×10^6 for OUA. For TFT in L5178Y TK$^+$/$-$ cells 10^6 can be plated in 9 cm dishes in 33 ml of medium. The relevant numbers for the surface growing Chinese hamster cells are for 6-TG up to 1.5×10^3 cells per cm^2 and for OUA up to 3×10^3 cells per cm^2, giving, in a 9 cm petri dish, 1×10^5 for 6-TG and 2×10^5 for OUA. For these cells, while the OUA system, unlike the 6-TG system, is not influenced to the same extent by metabolic co-operation, dead cells tend not to come off the surface thus making the discrimination of individual clones difficult at higher cell densities.

Important practical implications arise from the observed spontaneous mutant frequencies, from the limitations on what can be achieved at the biological level, and from the desirability of avoiding zero counts in all plates in a set. It is easy to see from the previous paragraphs that the TFT system because of its high spontaneous mutant frequency and high permissible density of cells must appear the most satisfactory assay system from the statistical point of view and gives further support for a preference for the suspension cell type. It is also clear that an unacceptably large

Table 3.1. *The probability that for a given mutant frequency all plates have zero mutants*

% cloning efficiency at expression time	mutant frequency *MF* per 10^6	*n* = number of plates per point 10^6 cells per plate						
		2	3	4	5	6	7	8
100	50	all less than 10^{-40}						
	5	all less than 10^{-4}						
	0.5	0.368	0.223	0.135	0.082	0.050	0.030	0.018
	0.05	0.905	0.861	0.819	0.779	0.741	0.705	0.670
50	50	all less than 10^{-20}						
	5	0.007	all less than 10^{-3}					
	0.5	0.607	0.472	0.368	0.287	0.223	0.174	0.135
	0.05	0.951	0.928	0.905	0.883	0.861	0.840	0.819
10	50	all less than 10^{-4}						
	5	0.368	0.223	0.135	0.082	0.050	0.030	0.018
	0.5	0.905	0.861	0.819	0.779	0.741	0.705	0.670
	0.05	0.990	0.985	0.980	0.975	0.970	0.966	0.961

Table 3.2. *The number of cells per plate necessary to ensure that a complete set of zero counts occur with a probability of less than 0.05*

Mutant Frequency *MF* per 10^6	*n* = number of plates per point						
	2	3	4	5	6	7	8
50	3×10^4	2×10^4	1.5×10^4	1.2×10^4	10^4	8.6×10^3	7.5×10^3
5	3×10^5	2×10^5	1.5×10^5	1.2×10^5	10^5	8.6×10^4	7.5×10^4
0.5	3×10^6	2×10^6	1.5×10^6	1.2×10^6	10^6	8.6×10^5	7.5×10^5

number of plates are requried for the OUA and 6-TG systems for Chinese hamster cells but that the 6-TG system is acceptable for L5178Y cells. As a consequence of the relative ease with which more cells may be treated and sampled, sufficient cells can be handled to justify continued work with the L5178Y OUA system. Clearly, more development work will be necessary with Chinese hamster cells perhaps by use of methodology to avoid sub-culture during expression time (O'Neill & Hsie, 1979) or by adapting them to grow in suspension.

3.2.3 The need for replicate cultures

There are two obvious sources of variation in any biological experiment of this type. Some variation exists in the number of cells distributed between plates from an individual culture, and this can be measured in terms of the colonies observed in different plates. However, there is another component of variation which is likely to be much larger than this, namely the variation between replicate cultures undergoing the same treatment in the same experiment. Reasons for this greater variability will be complex, but one important component must arise from the subculture of treated samples which is necessary to allow time for the expression of induced mutants (Leong *et al.*, 1985). Apart from this work, very few data are available, but one well-replicated study produced a between-culture variance about four times the size of the within-culture variance. Experience in many other fields of biological experimentation leads us to expect that this situation will normally obtain.

Any statistical analysis based only on the within-culture variation will therefore be likely to be spuriously sensitive and yield false significances. Replication of cultures within every experiment is essential in order to estimate both components of variation so that they can be combined properly for use in analysis.

This between-replicate variation is distinct from the variation of historical means; such historical data contain systematic variation between the times and places where experiments were done, systematic differences between operators and cultures, and between experimental techniques used. Historical data would usually be expected to give an over-estimate of the variation to be found in a single-centre, properly replicated but adequately controlled experiment. Therefore the between-replicates variation must be estimated in every experiment, although, as indicated in Section 3.4.6, it may be possible in special circumstances to combine historical estimates of between replicates variance.

It should be mentioned that subculturing will primarily affect the variation between replicates in numbers of colonies on mutation plates. Since cell number is determined at the time of plating, numbers of colonies on survival plates will not be subject to subculturing error, although they may instead be subject to dilution and counting errors. The importance of these effects may be less serious for bacterial 'treat and plate' assays and for yeast assays, but the same requirement for genuine replicates applies.

3.2.4 Accuracy of dilution

The analyses described below are performed as if all dilutions are completely accurate, so that the exact number of cells plated is known. In practice, errors will be introduced when performing dilutions, and, if these are excessive, the accuracy of the experiment will be seriously reduced. In the protocol which we advocate, where all cultures are replicated, the error introduced by dilution is in fact allowed for, but is compounded with other factors in the between-replicates error.

3.2.5 Repeating experiments

Since the practical difficulties in performing these mammalian cell assays limit the number of cultures and compounds within any one experiment, the reliability of estimates of variability will, inevitably, be similarly reduced. Thus, in addition to the requirement to confirm all data in independent experiments, repeat experiments are necessary to improve our estimates of between-replicates error (see Section 3.4.5).

3.3. METHODS OF ANALYSIS

3.3.1 Overview

A number of alternative methods have been proposed for the analysis of data from mammalian cell gene mutation assays (Clive *et al.*, 1979; Amacher *et al.*, 1980; Carver *et al.*, 1980; Hsie *et al.*, 1980, 1981; Snee & Irr, 1981; Irr & Snee, 1982; Boyd, 1982). From the statistical point of view all analyses are of mutant frequencies corrected for toxicity. Therefore the comments made should be considered as applying to all assays and may be extended to similar assays with bacteria and yeasts.

It is important to recognise several things about the use of statistical methodology. First, the responsibility for the appropriateness of particular methods in given situations rests with the user of those methods. Whether a particular method is appropriate or not depends on how the assay was done, what statistical assumptions can be made about the data, and the inferences to be made from the data. Secondly, the criteria for deciding whether or not a test substance is mutagenic in a given system are not only statistical. For the standard mammalian cell gene mutation assay the conventional criteria which have been used in decision-making are listed below and will be discussed in more detail subsequently.

(a) The negative control mutant frequency values adjusted for cell survival should lie within reasonable limits based on available data.

(b) The positive control mutant frequency values adjusted for cell survival should lie within reasonable limits based on available data.

(c) The test substance should display mutant frequency values greater than those of the negative control allied to the existence of a relationship with dose. This relationship may be non-linear, and with some chemicals may indeed be decreasing over part of the dose range.

(d) Some threshold value of adjusted mutant frequency may be required to have been attained at one dose at least. The toxicity of the test substance should be taken into consideration, with only doses showing cell survival above a certain percentage being considered (usually 10%).

(e) The result (whether positive or not) should be reproducible in at least two independent experiments.

Thus, in addition to assessing whether the test substance causes increased mutant frequencies and whether a dose relationship exists (criterion c), the Negative and Positive controls (criteria a and b) are required in order to assess whether or not a reasonable experiment has been done. A threshold value (criterion d) may be used to assess whether any produced (or likely to be produced) increases in mutant frequency are biologically important. With a threshold, it is necessary to consider the toxicity of the test substance. For a chemical producing weak increases in mutant frequency, exceeding a given threshold may only be achieved with a dose that kills all the cells.

Statistical methodology and ideas are involved in framing these criteria and we have to consider how to apply statistical analyses to establish whether data satisfy them.

3.3.2 Negative and positive controls

When a test substance is assayed, the negative and positive control means for mutant frequency for that experiment should be compared with historical values and if they do not lie within a reasonable range, rejection of the experimental data should be considered. These control means should also be comparable with values available from sources such as journals and colleagues in other laboratories. At its simplest this is standard quality control and can be treated as such, with reasonable ranges being constructed from available data. The values from the current experiment are then checked to ensure that they lie within these intervals. Standard 'reasonable ranges' are known as confidence intervals. It is usual to use 95% and 99% confidence intervals as 'warning' and 'action' levels respectively. The 95% confidence interval is the range within which one is 95% certain that an estimate of the control mean mutant frequency from a valid individual experiment should lie. Similarly for the 99% confidence interval. If sufficient data are available these intervals can be calculated simply as those values which have 2.5% of the data above and below them (i.e. 95% between them) for the 95% confidence interval and 0.5% above and below for the 99% interval. If insufficient data are available these intervals can be calculated as: mean value $\pm$ a Student's t-value $\times$ standard deviation.

Box 3.1 shows the procedure for calculating these confidence intervals. A standard procedure is to: (a) accept the experiment if both control means lie within the 95% confidence interval; (b) accept the experiment if either or both control means lie outside the 95% interval but within the 99% interval but consider this as a warning that something might have gone wrong and review the experimental procedure; and (c) reject the experiment if either or both control means lie outside the 99% confidence interval. This procedure is a convenient approximation. The distribution of spontaneous mutants in starting cultures is extremely complex (Lea & Coulson, 1949) and skewed so that the test may well give a rejection rate higher than the expected approximate 2%, with an excess of higher negative control mutant frequencies. Since an excessive negative control mutant frequency mean will reduce the sensitivity of the assay, it is appropriate to reject such values even though they may have arisen by chance. If an experiment is rejected, investigations may need to be undertaken as to the cause of the extreme values.

3.3.3 Published methods

Evaluating whether the test substance induces increases in mutant frequency and whether a dose relationship exists is central to deciding if it is mutagenic or not in a given experiment. In much of the discussion in the literature on the evaluation of mutagenicity test data these results became intertwined with the question of attaining a threshold value, but to a large extent they are separate issues. Methods proposed for looking at dose relationships fall into the two general statistical categories of linear regression and non-parametric tests for ordered alternatives. The regression methods make certain assumptions about the distribution of mutant frequency values and attempt to fit a function of these to a function of dose. Hsie & co-workers (Tan & Hsie, 1981; Hsie *et al.*, 1980, 1981) propose weighted regression be used. This assumes that mutant frequency is normally distributed but that the variance of this distribution changes between doses. A linear regression line (i.e. a straight line) is then fitted to the mean mutant frequency value weighting by the reciprocal of some function representing the variance at that dose. Two functions are discussed here, the first being the empirical variance calculated from the plate counts, the second being a variance assuming plate counts have Poisson distribution. Unweighted regression assumes that the variance at all the doses is constant and weighting is a way of using regression methodology when this assumption does not hold.

Irr & Snee (1982) (also Snee & Irr, 1981) take a somewhat different stance. They say that the mutant frequency values are not normally

Box 3.1. *A simulated sample of mean adjusted mutant frequency values (MFs) for negative controls and their variances*

Experiment	Mean mutant frequency *MF*	Variance of *MF* *V*	Weight (1/*V*) *W*
1	1.18	0.119	8.403
2	1.01	0.201	4.975
3	0.97	0.105	9.524
4	0.91	0.099	10.101
5	1.20	0.130	7.692
6	1.05	0.110	9.091
7	0.99	0.109	9.174
8	1.22	0.132	7.576
9	1.07	0.107	9.346
10	1.01	0.101	9.901

The method of calculating *MF*, *V* and *W* from the experimental data is set out in Section 3.4.1 and Section 3.4.2.

$$\text{Weighted mean value } (WMV) = \frac{\sum(W \times MF)}{\sum W} = \frac{8.403 \times 1.18 + \ldots + 9.901 \times 1.01}{8.403 + \ldots + 9.901} = 1.05$$

$$\text{Weighted variance} = \frac{\sum W(MF - WMV)^2}{\sum W} = \frac{8.403(1.18 - 1.05)^2 + \ldots + 9.901(1.01 - 1.05)^2}{8.403 + \ldots + 9.901} = 0.00995$$

$$\text{Standard deviation} = \sqrt{(\text{variance})} = \sqrt{(0.00995)} = 0.100$$

Confidence intervals are

$$WMV \pm (\text{Student's } t \text{ value})(\text{Standard deviation})$$

From tables of Student's *t*-test, *t* for a probability of 0.95 with 9 degrees of freedom is 2.262. This gives intervals of

$$1.05 + (2.262 \times 0.100) = 1.28$$

and

$$1.05 - (2.262 \times 0.100) = 0.82$$

Similarly *t* for a probability of 0.99 with 9 degrees of freedom is 3.250. Therefore the 99% confidence intervals are

$$1.05 + (3.250 \times 0.100) = 1.38$$

and

Box 3.1. (*cont.*)

$$1.05-(3.250\times 0.100)=0.725$$

In this instance it would be considered a 'warning' if the *MF* for the negative control in the next experiment performed lay between 0.72 and 0.82, or between 1.28 and 1.38. The experiment would be rejected if the negative control *MF* was less than 0.72 or greater than 1.38.

distributed nor do they have equal variances, but they can be easily transformed into a variable that does have these properties. Applying well-known techniques (Box & Cox, 1964) to their data they recommend a log transformation (i.e. log transformation before analysis). Other workers (Clive *et al.*, 1979; Carver *et al.*, 1980) also propose regression techniques for evaluating the existence of a relationship between mutant frequency and dose although these authors have used functions of relative survival rather than dose against which to fit a function of mutant frequency. If there is a close linear relationship between cell survival and dose then it does not matter too much which approach is used. The argument used to support the use of a function of cell survival is that it enables one to see clearly the range of survival within which mutagenesis can be assessed (Newbold *et al.*, 1979).

Non-parametric tests for evaluating the existence of a dose relationship have been investigated by Boyd (1982) who considered and discussed three such tests. As non-parametric tests these make only a few general assumptions about the data. In particular, no distributional form (e.g. normal or Poisson) is assumed and there is no restriction to a particular function.

A number of other methods which fit into neither the category of regression nor that of non-parametric tests have been proposed for analysing these data. Amacher *et al.* (1980) ignore the dose relationships and cell survival aspects of the data and use a two-sample *t*-test on the log transformed mutant counts, not adjusted for cell survival, to compare controls with the pooled results for all doses of the test substance where survival is >20%. This approach has been criticised and there is correspondence about it between the groups of Amacher & Clive in Mutation Research (Amacher & Salsburg, 1981; Clive *et al.*, 1981; Clive & Hajian, 1981). Newbold *et al.* (1979) use two-sample *t*-tests to compare control mutant frequency with the mutant frequencies at the various doses of test substance. Finally, an interesting paper by Lee & Caspary (1983) describes efforts to mathematically model the biological processes involved in the L5178Y mouse lymphoma forward mutation assay.

3.3.4 Threshold values

If a relationship between dose and mutant frequency is established it is useful to review the data to see whether a certain threshold value has been attained, or could be attained, within the 100%–10% survival range. Suggestions as to what this threshold value should be have been published (Clive *et al.*, 1979). Much of the discussion about what this threshold value should be has been couched in statistical terms with the apparent aim of producing a simple rule that can substitute for more formal statistical tests. We suggest that threshold values should be used as an aid in interpreting an effect, rather than in determining whether or not an effect exists.

3.3.5 Recommended procedures

Five criteria are given for deciding whether a substance is mutagenic. The most important of these is whether there is an increase in mutant frequency over negative control levels and whether a dose relationship exists. The non-parametric approaches of Boyd (1982) have the advantage of few (if any) distributional assumptions being made and the disadvantage of low power with a larger number of results being required. In general, the assays with mammalian cells produce insufficient genuinely independent replicate observations for the use of non-parametric methods.

Parametric methods have the advantages of high power and the possibility of performing sophisticated analyses, but the disadvantage of requiring distributional assumptions to be made. We prefer weighted regression to transformation of the data because it allows a test for a direct relationship between mutant frequency and dose. The choice of form of weighting is less definite. One form involving variances calculated from the individual plate counts has the advantage of not requiring distributional assumptions to be made about these counts, but the disadvantage of using estimates of variance which are likely to be imprecise because of the small numbers of observations involved. An alternative weighting not involving calculated variances is more precise but an assumption of Poisson distributed counts is required. The evidence from available data for/against the reasonableness of assuming Poisson distributed counts was equivocal. Since Poisson-derived weights are simpler to calculate, and may give a more realistic weighting where plates have been lost, we prefer their use. It should be stressed that the choice of weighting is likely in practice to have little effect on the analysis.

Table 3.3. *Example data*

(a) Mutation frequency plates

Treatment	Concentration D	Cells plated P_m	Numbers plates N_m	Plate counts (m)				Mean $\bar{m}$
Control	0.0	10^6	4	2	1	1	0	1.0
Control	0.0	10^6	4	0	2	2	1	1.25
Test substance	5.0	10^6	4	3	4	1	2	2.5
Test substance	5.0	10^6	4	5	4	3	4	4.0
Test substance	10.0	10^6	4	7	6	5	8	6.5
Test substance	10.0	10^6	4	6	6	3	7	5.5
Test substance	15.0	10^6	4	9	10	8	11	9.5
Test substance	15.0	10^6	4	10	11	11	9	10.25
+Control	25.0	10^6	4	27	28	30	27	28.0
+Control	25.0	10^6	4	29	33	31	32	31.25

(b) Survival plates

Treatment	Concentration D	Cells plated P_c	Numbers plates N_c	Plate counts (c)		Mean $\bar{c}$
Control	0.0	200	2	183	189	186.0
Control	0.0	200	2	201	190	195.5
Test substance	5.0	200	2	156	161	158.5
Test substance	5.0	200	2	160	159	159.5
Test substance	10.0	200	2	120	117	118.5
Test substance	10.0	200	2	113	122	117.5
Test substance	15.0	200	2	81	83	82.0
Test substance	15.0	200	2	82	81	81.5
+Control	25.0	200	2	197	203	200.0
+Control	25.0	200	2	201	202	201.5

3.3.6 Terminology

Earlier in this section regression methods were mentioned but in the next section (Section 3.4) only analysis of variance tables are displayed. Regression and analysis of variance are highly interrelated techniques and so it is possible to incorporate regression calculations into an analysis of variance. This is, in fact, what was done, and the Linear Relationship (Lin. rel) term in the tables of Section 3.4 is the appropriate term for regression of *MF* on dose.

3.4 EXAMPLE DATA AND CALCULATIONS

The data in Table 3.3. have been constructed to illustrate calculations for testing whether a test substance produces an increase in mutant frequency and whether a dose relationship exists between mutant frequency and dose of test substance. It is a data set with two replicate cultures for each of one negative control, one positive control and three doses of a test substance. Two cloning efficiency plates, each with 200 cells, and four mutant frequency plates, each with 10^6 cells, have been assumed. It has also been assumed that mutant frequency is to be expressed as mutants per 10^6 cells.

In the examples, values are calculated to at least three decimal places, but in practice the full precision of the calculator or computer should be used. This is important because substantial rounding errors may easily be introduced.

3.4.1 Calculation of mutant frequency (*MF*) and its weight (*W*)

As discussed in Section 3.3, the weighted analysis described here can be performed using either Poisson-derived or empirically derived weights. Both methods are given here and both should give closely similar end results.

3.4.1.1 *Calculation using Poisson-derived weights*

In this calculation it is assumed that numbers of colonies on plates are distributed according to the Poisson distribution. Since the variance of the Poisson distribution is equal to the mean, it is not necessary to calculate variances of the actual plate counts. It must be emphasised that the raw plate counts and not the mutant frequencies are assumed to be Poisson distributed, the distribution of the calculated mutant frequencies will certainly not be Poisson.

First calculate the mean colonies per plate ($\overline{m}$) for mutation plates for each replicate treatment in Table 3.3:

$$\overline{m} = \frac{\sum m}{N_m}$$

where N_m is the number of mutation plates for that replicate treatment. For Dose 0 replicate 1

$$\overline{m} = \frac{2+1+1+0}{4} = 1.0$$

The means for the survival plates ($\overline{c}$) are calculated in the same way as those for the mutation plates.

Next calculate the dilution factor K which is a function of P_c and P_m (the cells per plate). Since these are constant throughout these data K is a constant.

$$K = \frac{P_c \times 10^6}{P_m} = \frac{200 \times 10^6}{10^6} = 200$$

Now the mutant frequency can be calculated as

$$MF = \text{K} \times \frac{\overline{m}}{\overline{c}}$$

i.e. for Dose 0 replicate 1

$$MF = 200 \times \frac{1.0}{186} = 1.075$$

The variance of the mutant frequency is obtained by evaluating the general variance formula for a function (see, for example, Topping, 1972). It is also basically the formula given in Clive *et al.* (1979) without their variance of dilution. Since, however, the plate counts are assumed to be Poisson distributed, the mean plate count is used as an estimate of variance.

$$V = \frac{K^2}{\overline{c}^2}\left[\frac{\overline{m}}{N_m} + \frac{\overline{m}^2}{N_c \overline{c}}\right]$$

i.e. for Dose 0 Replicate 1

$$V = \frac{200^2}{186^2}\left[\frac{1.0}{4} + \frac{1.0^2}{2 \times 186}\right]$$
$$= 0.292$$

Using the variance of the MF a weight (W) can be calculated where $W = 1/V$ $= 3.425$ for Dose 0 replicate 1

Performing these calculations produces the weights and mutant frequencies shown in Box 3.2.

3.4.1.2 *Calculation using empirical weights*

The variance of the mutant frequency can also be estimated from the actual plate counts. For this, the variance of each set of survival and mutation plate counts for each replicate is required. It is calculated for the mutation plates as

$$Var_m = \frac{\sum(m - \overline{m})^2}{N_m - 1}$$

i.e. for Dose 0 replicate 1 mutation plates

$$Var_m = \frac{(2-1)^2 + (1-1)^2 + (1-1)^2 + (0-1)^2}{4-1}$$
$$= 0.667$$

Box 3.2. *Calculation of mutant frequency and Poisson-based weight for each replicate*

Treatment	Dose	Surv. mean $\bar{c}$	Number of plates N_c	Mutation mean $\bar{m}$	Number of plates N_m	Dilution factor K	Variance V	Weight W	Mutant frequency MF
Control	0	186.0	2	1.0	4	200	0.292	3.423	1.075
Control	0	195.5	2	1.25	4	200	0.331	3.019	1.279
Treated	5	158.5	2	2.5	4	200	1.027	0.974	3.155
Treated	5	159.5	2	4.0	4	200	1.651	0.606	5.016
Treated	10	118.5	2	6.5	4	200	5.137	0.195	10.970
Treated	10	117.5	2	5.5	4	200	4.357	0.230	9.362
Treated	15	82.0	2	9.5	4	200	17.402	0.0575	23.171
Treated	15	81.5	2	10.25	4	200	19.313	0.0518	25.153
+ Control	(25)	200.0	2	28.0	4	200	8.960	0.112	28.000
+ Control	(25)	201.5	2	31.25	4	200	10.084	0.0992	31.017

Note:
Where:

$$K = \frac{P_c \times 10^6}{P_m}$$

$$V = \frac{K^2}{\bar{c}^2}\left[\frac{\bar{m}}{N_m} + \frac{\bar{m}^2}{N_c \bar{c}}\right]$$

$W = 1/V$ ________ calculated *before* V rounded down to three decimal places.

$$MF = K \times \frac{\bar{m}}{\bar{c}}$$

(note that MF is mutant frequency $\times 10^6$)

and for survival plates

$$Var_c = \frac{(183-186)^2+(189-186)^2}{2-1} = 18.0$$

In this instance the actual variances of the plate counts are used in estimating the variance of the mutant frequency.

$$V = \frac{K^2}{\bar{c}^2}\left[\frac{Var_m}{N_m} + \frac{(\bar{m})^2 Var_c}{N_c(\bar{c})^2}\right]$$

i.e. for Dose 0 replicate 1

$$= \frac{200^2}{186^2} \times \left[\frac{0.667}{4} + \frac{1.0^2 \times 18.0}{2 \times 186^2}\right]$$

$$1.156 \times 0.167$$

$$= 0.193$$

and

$$W = \frac{1}{V} = 5.180$$

The mutant frequency, *MF*, is calculated as in Section 3.4.1.1.

3.4.2 Weighted analysis of variance

Using the values in Box 3.2, a weighted analysis of variance (regression) as suggested by Hsie & co-workers (Tan & Hsie, 1981; Hsie *et al.*, 1980, 1981) can now be performed.

3.4.2.1 *Calculation of means and sums of squares*

First of all some weighted means need to be calculated. The overall mean is:

$$\text{Overall mean} = \frac{\sum(W \times MF)}{\sum(W)}$$

where

$$\begin{aligned}\sum(W \times MF) = & (3.423 \times 1.075) + (3.019 \times 1.279) + (0.974 \times 3.155) \\ & + (0.606 \times 5.016) + (0.195 \times 10.970) + (0.230 \times 9.362) \\ & + (0.0575 \times 23.171) + (0.0518 \times 25.153) \\ & + (0.112 \times 28.000) + (0.0992 \times 31.017) \\ = & 26.794\end{aligned}$$

$$\begin{aligned}\sum(W) = & 3.423 + 3.019 + 0.974 + 0.606 + 0.195 \\ & + 0.230 + 0.0575 + 0.0518 + 0.112 + 0.0992 \\ = & 8.768\end{aligned}$$

and

$$\frac{\sum(W \times MF)}{\sum(W)} = \frac{26.794}{8.768} = 3.056$$

The Test mean (mean of all treatments except the positive control) is calculated similarly

$$\begin{aligned}\sum(W \times MF) &= (3.423 \times 1.075) + (3.019 \times 1.279) + (0.974 \times 3.155) \\ &\quad + (0.606 \times 5.016) + (0.195 \times 10.970) + (0.230 \times 9.362) \\ &\quad + (0.0575 \times 23.171) + (0.0518 \times 25.153) \\ &= 20.581\end{aligned}$$

$$\begin{aligned}\sum(W) &= 3.423 + 3.019 + 0.974 + 0.606 + 0.195 \\ &\quad + 0.230 + 0.0575 + 0.0518 \\ &= 8.556\end{aligned}$$

and

$$\frac{\sum(W \times MF)}{\sum(W)} = \frac{20.581}{8.556} = 2.405$$

The means for the individual treatments are given by similar summations but only over the values for that treatment, i.e.

$$\text{Control mean} = \frac{(3.423 \times 1.075) + (3.019 \times 1.279)}{3.423 + 3.019} = 1.171$$

$$\text{Test substance mean first dose} = \frac{(0.974 \times 3.155) + (0.606 \times 5.016)}{0.974 + 0.606} = 3.869$$

$$\text{Test substance mean second dose} = \frac{(0.195 \times 10.970) + (0.230 \times 9.362)}{0.195 + 0.230} = 10.100$$

$$\text{Test substance mean third dose} = \frac{(0.0575 \times 23.171) + (0.0518 \times 25.153)}{0.0575 + 0.0518} = 24.110$$

$$\text{Positive control mean} = \frac{(0.112 \times 28.000) + (0.0992 \times 31.017)}{0.112 + 0.0992} = 29.417$$

The various sums of squares can now be calculated and all are of the form

$$\sum[(MF - \text{mean } MF)^2 \times W].$$

$$\begin{aligned}\text{Total sum of squares} &= (1.075 - 3.056)^2 3.423 + (1.279 - 3.056)^2 3.019 \\ &\quad + (3.155 - 3.056)^2 0.974 + (5.016 - 3.056)^2 0.606 \\ &\quad + (10.970 - 3.056)^2 0.195 + (9.362 - 3.056)^2 0.230 \\ &\quad + (23.171 - 3.056)^2 0.0575 + (25.153 - 3.056)^2 0.0518 \\ &\quad + (28.000 - 3.056)^2 0.112 + (31.017 - 3.056)^2 0.0992 \\ &= 242.464\end{aligned}$$

Next the sum of squares for differences between individual treatments is calculated. For each individual treatment the weights are summed and the MF value used is the mean for that treatment, i.e.

Between individual
treatment sum
of squares $=(1.171-3.056)^2(3.423+3.019)$
$+(3.869-3.056)^2(0.974+0.606)$
$+(10.100-3.056)^2(0.195+0.230)$
$+(24.110-3.056)^2(0.0575+0.0518)$
$+(29.417-3.056)^2(0.112+0.0992)$
$=240.235$

The next sum of squares classifies the positive control as one group and the remaining treatments as a second group. As above, the mean MF of all treatments within the group is used and the weights of all the treatments summed.

Between Groups
sum of squares $=(29.417-3.056)^2(0.112+0.0992)$
$+(2.405-3.056)^2(3.423+3.019+0.974+0.606+$
$0.195+0.230+0.0575+0.0518)$
$=150.390$

To calculate the linear relationship for test substance sum of squares two other sums of squares are required. To obtain these, the weighted mean dose for the test substance is first calculated as

$$\text{Weighted mean dose}=\frac{\sum(\text{Dose}\times W)}{\sum(W)}$$

$$\sum(\text{Dose}\times W)=0(3.423+3.019)+5(0.974+0.606)$$
$$+10(0.195+0.230)+15(0.0575+0.0518)$$
$$=13.790$$

$$\sum(W)=3.423+3.019+0.974+0.606$$
$$+0.195+0.230+0.0575+0.0518$$
$$=8.556$$

$$\frac{\sum(\text{Dose}\times W)}{\sum(W)}=\frac{13.790}{8.556}=1.612$$

The first of the sums of squares needed is that for doses of test substance.

$$\text{Doses sum of squares}=\sum(\text{Dose}-\text{mean dose})^2\times W$$
$$=(0.0-1.612)^2(3.423+3.019)+(5.0-1.612)^2(0.974+0.606)$$
$$+(10.0-1.612)^2(0.195+0.230)+(15.0-1.612)^2(0.0575+0.0518)$$
$$=84.370$$

The second is a cross-product of mutant frequency (MF) with dose. Note that in this case the mean mutant frequency used is that excluding the positive control.

$$\begin{aligned}\text{Cross-product } MF \text{ and dose} &= \sum[(MF - \text{mean } MF)(\text{dose} - \text{mean dose})W] \\ &= (1.075-2.405)(0-1.612)3.423+(1.279-2.405)(0-1.612)3.019 \\ &\quad +(3.155-2.405)(5-1.612)0.974+(5.016-2.405)(5-1.612)0.606 \\ &\quad +(10.970-2.405)(10-1.612)0.195+(9.362-2.405)(10-1.612)0.230 \\ &\quad +(23.171-2.405)(15-1.612)0.0575 \\ &\quad +(25.153-2.405)(15.0-1.612)0.0518 \\ &= 79.847\end{aligned}$$

Now the sum of squares for the linear relationship can be calculated.

$$\text{Linear relation sum of squares} = \frac{(\text{Cross-product } MF \text{ and dose})^2}{\text{Doses sum of squares}} = \frac{(79.847)^2}{84.370} = 75.566$$

3.4.2.2 *Setting up the analysis of variance table*

The above values can be used to set up the analysis of variance table shown in Table 3.4.

The Total sum of squares is 242.464 and, since ten mutant frequencies are compared, it has $10-1=9$ degrees of freedom.

The Between Treatments sum of squares is 240.235 with, since five treatments are compared, $5-1=4$ degrees of freedom.

The Residual Error sum of squares, which in this analysis corresponds to the Between Replicates error sum of squares, is $242.464-240.235=2.229$ with $9-4=5$ degrees of freedom.

The Between Treatments error can be further split into the Between Groups (positive control versus rest) sum of squares which is 150.390 with $2-1=1$ degree of freedom and the sum of squares for the Linear Relationship of the Test Substance with dose which is 75.566 with 1 degree of freedom.

This leaves a sum of squares of
$240.235-150.390-75.566=14.279$ with $4-1-1=2$
degrees of freedom corresponding to the non-linear component of the Test Substance dose response.

The mean squares in Table 3.4 are obtained by dividing each sum of squares by the appropriate number of degrees of freedom. The ratio of each mean square to the Residual (Between Replicates) Error mean square is calculated. These variance ratios are tested against critical values in F (variance ratio) tables, where n_1 = degrees of freedom of the item being

Table 3.4. *Analysis of variance table for data from Box 3.2.*

Item	Degrees of freedom	Sum of squares	Mean square	F	Probability
Between treatments	4	240.235	60.059	134.661	$P<0.001$
Between groups	1	150.390	150.390	337.197	$P<0.001$
Linear relation	1	75.566	75.566	169.430	$P<0.001$
Non-linear component	2	14.279	7.140	16.009	$P<0.01$
Between replicates	5	2.229	0.446		
Total	9	242.464			

Note: that 0.446 is the Between Replicates Error Mean Square used in subsequent calculations.

tested and n_2 = degrees of freedom of the residual error.

It can be seen that there are highly significant differences between individual treatments ($F=134.66$, $n_1=4$, $n_2=5$, $P<0.001$ critical value = 31.09). The method of determining which treatments are significantly different from the negative control is described in Section 3.4.3. There is also a highly significant linear trend ($F=169.430$, $n_1=1$, $n_2=5$, $P<0.001$ critical value = 47.18). The method for calculating a slope and its variance are outlined in Section 3.4.4, as are the modifications to these calculations where the negative control is to be excluded from a test for trend.

3.4.3 Comparison between individual treatments

Although Table 3.4 indicates that mutant frequencies from different treatments are significantly more variable than mutant frequencies of replicates of the same treatment, it is necessary to test whether the mutant frequency at a particular dose level is significantly higher than that in the negative control. The procedure for doing this uses the estimates of the between replicates residual error mean square (*EMS*) from Table 3.4 and the weights from Box 3.2 to obtain estimates of the variance and standard error of the mean mutant frequency at each dose. To calculate the Variance of the mean mutant frequency at Dose 0:

$$Var_0 = \frac{EMS}{\text{SUM}[W_0]} = \frac{0.446}{3.423 + 3.019} = 0.069$$

Results for the remaining doses are given in Box 3.3.

The mutant frequency for each dose of the test agent can be compared with that of the negative control by means of a t-test. For instance for Dose 5.0

$$t = \frac{(\bar{m}_5 - \bar{m}_0)}{\sqrt{(Var_5 + Var_0)}} = \frac{(3.869 - 1.171)}{\sqrt{(0.282 + 0.069)}} = 4.553$$

The calculated values of t for the comparison of each dose with the negative control are also given in Box 3.3. The degrees of freedom for t are the same as those for EMS, the between replicates error mean square (Table 3.4), in this instance 5. It should be noted that in this example, the t-test is a two-sided test of significance, since it is testing for both increases and decreases in mutant frequency in the treated sample. Normally only an increase in mutant frequency is of interest, and a one-sided test of significance may be considered. The grounds for choosing between one- and two-sided tests are discussed in Chapter 1. If desired, the t-test can be treated as a one-sided test using the same statistical tables, by testing only for an increase in mutant frequency, and by dividing by two the probability level stated in the table. Because mutant frequencies at several doses are compared to the same negative control, some form of allowance must be made for multiple comparisons, for instance by using Dunnett's test (Dunnett, 1955), or the Bonferroni correction (see Chapter 1).

We recommend, however, that, rather than try to determine an exact estimate of probability in a borderline case, it will be far more useful to perform an additional experiment. Methods for combining results from repeat experiments are given in Section 3.4.5 below.

3.4.4 Test of linear trend

3.4.4.1 *Test including negative control*

The analysis of variance in Box 3.4 shows a highly significant linear trend. Further analysis is possible. From the cross-product of Mutant frequency, *MF*, and dose and the Doses sum of squares, the slope, b, of the straight line can be calculated.

$$\text{Slope, } b = \frac{\text{Cross-product } MF \text{ and dose}}{\text{Doses sum of squares}} = \frac{79.847}{84.370} = 0.946$$

Therefore the straight line relationship between Mutant frequency and Dose for the test substance is one with a slope of 0.946 going through the

Box 3.3. *Comparison of mutant frequency at different doses of test substance with negative control*

Treatment	Dose D	Mean mutant frequency MF	Sum Wts	Variance mean MF	Diff	Var_{diff}	t	P
Negative control	0.0	1.171	6.442	0.0692	—			
Test substance	5.0	3.869	1.580	0.282	2.698	0.351	4.553	<0.01
Test substance	10.0	10.100	0.425	1.049	8.929	1.118	8.444	<0.001
Test substance	15.0	24.110	0.109	4.092	22.939	4.161	11.245	<0.001
+Control	25.0	29.417	0.211	2.114	28.246	2.183	19.117	<0.001

Notes:
Error Mean Square (EMS) from Table 3.4 = 0.446, with 5 degrees of freedom so that t has 5 degrees of freedom.

$$\text{Variance of mean } MF = \frac{EMS}{\text{Sum Wts for that } MF}$$

(note that MF is mutant frequency $\times 10^6$)

point 2.405, 1.612 (see also Section 3.4.2.1) on the Mutant frequency and Dose scales respectively. With this information, the intercept a, which is the value of Mutant frequency when Dose is zero, can be calculated as

$$\begin{aligned} a &= (\text{mean } MF) - \text{slope (mean Dose)} \\ &= 2.405 - (0.946 \times 1.612) \\ &= 0.880 \end{aligned}$$

So that the model is

$$\begin{aligned} MF &= a + b \text{ Dose} \\ &= 0.880 + 0.946 \text{ Dose} \end{aligned}$$

The variance of this estimate of the slope is

$$\begin{aligned} Var_b &= \frac{EMS}{\text{Doses sum of squares}} \\ &= \frac{0.446}{84.370} \\ &= 0.005 \end{aligned}$$

The estimate of slope and its variance are required for comparing and combining repeat experiments as described in Section 3.4.5.

3.4.4.2 *Test excluding negative control*

A review of the linear relationship calculations so far will show that they have been performed assuming the negative control to be the zero treatment of test substance, an alternative view being that the negative control data should be excluded from the test for linear trend. The arguments for and against these alternatives are presented in the Discussion Section 3.5. To aid understanding of these arguments the calculations have been repeated excluding the negative control from the estimate of linear trend. For this analysis, the negative control, the positive control and the test substance results are considered as three separate groups, and the following values require to be recalculated.

The weighted mean of the test group is now

$$\text{Test group mean } MF = \frac{(0.974 \times 3.155) + (0.606 \times 5.016) + (0.195 \times 10.970) + (0.230 \times 9.362) + (0.0575 \times 23.171) + (0.0518 \times 25.153)}{0.974 + 0.606 + 0.195 + 0.230 + 0.0575 + 0.0518}$$

$$= 6.168$$

and

$$\text{Weighted mean Dose} = \frac{5(0.974 + 0.606) + 10(0.195 + 0.230) + 15(0.0575 + 0.0518)}{0.974 + 0.606 + 0.195 + 0.230 + 0.0575 + 0.0518} = 6.522$$

giving

$$\text{Doses sum of squares} = (5-6.522)^2(0.974+0.606)+(10-6.522)^2(0.195+0.230) +(15-6.522)^2(0.0575+0.0518) = 16.657$$

and

$$\begin{aligned}\text{Cross-product } MF \text{ and Dose} = &(3.155-6.168)(5-6.522)0.974\\ &+(5.016-6.168)(5-6.522)0.606\\ &+(10.970-6.168)(10-6.522)0.195\\ &+(9.362-6.168)(10-6.522)0.230\\ &+(23.171-6.168)(15-6.522)0.0575\\ &+(25.153-6.168)(15-6.522)0.0518\\ &=27.967\end{aligned}$$

and

$$\text{Linear Relation sum of squares} = \frac{27.967^2}{16.657} = 46.956$$

The Between Groups sum of squares is now

$$\begin{aligned}&=(1.171-3.056)^2(3.423+3.019)\\ &\quad+(6.168-3.056)^2(0.974+0.606+0.195+0.230+0.0575+0.0518)\\ &\quad+(29.417-3.056)^2(0.112+0.0992)\\ &=190.129\end{aligned}$$

The various sums of squares and mean squares are calculated as for Table 3.4. There are, however, now three groups giving Between Groups 3–1 = 2 degrees of freedom and as a result one less degree of freedom for the non-linear error component. The resulting analysis of variance is set out in Table 3.5.

Clearly, there are highly significant differences between the negative control, positive control and average response at all doses of test substance and it can be shown by the t-test procedure outlined in Section 3.4.3 that the positive control and the average response over all doses are both significantly higher than the negative control. It can also be seen, with the model data tested here, that omission of the negative control reduces the significance of the non-linear component.

3.4.5 Comparisons between experiments, and combining information from independent experiments

Where, as we have recommended, an experiment has been repeated, it is important to be able to determine whether a result is

Table 3.5. *Analysis of variance table, excluding the negative control from the test for trend.*

Item	Degrees of freedom	Sum of squares	Mean square	F	Probability
Between treatments	4	240.235	60.059	134.661	$P<0.001$
Between groups	2	190.129	95.065	213.150	$P<0.001$
Linear relation	1	46.956	46.956	105.283	$P<0.001$
Non-linear component	1	3.15	3.15	7.063	$P<0.05$
Between replicates	5	2.229	0.446		
Total	9	242.464			

consistent between two or more experiments, and, where it *is* consistent, to be able to combine the experiments. The procedure for comparison of mutant frequency at an individual dose versus the negative control can readily be used to compare induced mutant frequencies at the same dose in separate experiments. We suggest that, rather than comparing absolute mutant frequencies at each dose, it is more appropriate to use the difference between the mutant frequency at the dose under consideration and the control mutant frequency. This difference, and its variance, have already been calculated in Section 3.4.3. Using the model values from dose level 5.0, Box 3.3 and additional data from Box 3.4 as an example, analysis of variance (Box 3.5) can be performed.

Firstly, weights are calculated from each variance

$$W_{diff}=\frac{EMS}{Var_{diff}}$$

Next a combined weighted mean difference (CD) must be calculated

$$CD=\frac{\sum(diff\times W_{diff})}{\sum(W_{diff})}$$

$$=\frac{(2.698\times 1.271)+(2.013\times 0.909)+(3.829\times 0.535)}{1.271+0.909+0.535}$$

$$=2.692$$

The between experiments sum of squares is calculated as $\sum(diff-CD)^2 W_{diff}$

Box 3.4. *Simulated data for testing consistency between experiments and for combining experiments*

		Experiment 1	Experiment 2	Experiment 3
Difference between dose 5 and control	*diff*	2.698	2.013	3.829
Variance	var_{diff}	0.351	0.550	0.977
Weight	W_{diff}	1.271	0.909	0.535
Error Mean Square	*EMS*	0.446	0.500	0.523
Degrees of freedom	*d.f.*	5	5	6

Box 3.5. *Test for consistency between experiments*

	d.f.	Sum of squares	Mean square	*F*	Probability
Between experiments	2	1.111	0.555	1.058	*NS*
Within experiments	16	7.867	0.525		

$$= (2.698 - 2.692)^2 1.271 + (2.013 - 2.692)^2 0.909 + (3.829 - 2.692)^2 0.535$$
$$= 1.111$$

with the number of degrees of freedom one fewer than the number of experiments

$$= 3 - 1 = 2$$

The within experiments sum of squares is obtained by pooling the information from the individual experiments

$$\sum(EMS d.f.)$$
$$= 0.446 \times 5 + 0.500 \times 5 + 0.523 \times 6$$
$$= 7.867$$

with $\sum(d.f.) = 5 + 5 + 6 = 16$ degrees of freedom

This enables the analysis of variance set out in Box 3.5 to be constructed and a test of the consistency of the result between experiments to be made. The resulting variance ratio, 1.058 is less than the 5% critical *F* value, 3.63

for 2 and 16 degrees of freedom. We therefore conclude that the three experiments are consistent. It should be noted that this is a two-sided test, since a repeat experiment may give either a higher or lower value than the first. The Combined Within Experiments Mean square ($CEMS=0.525$) will be used in testing the significance of the combined difference.

Since the experiments are consistent, the combined mean difference can be tested for significance in exactly the same way as the difference between a treatment and a control mutant frequency in an individual experiment. As in that instance, the variance of the combined mean difference (CD) is the Combined Within Experiments Mean Square ($CEMS=0.525$) divided by the sum of the weights.

$$t=\frac{CD}{\surd(CEMS/\sum(W_{diff}))}$$

$$=\frac{2.683}{\surd(0.525/(2.849+1.818+1.024))}$$

$$=8.834$$

with $\sum(d.f.)=16$ degrees of freedom (the 0.1% critical level for 16 degrees of freedom is 4.015).

Exactly the same procedure can be used to test dose-response slopes from different experiments for consistency, and to perform a combined test of significance. The estimates of slope and their variances from different experiments, as obtained in Section 3.4.4.1, are treated in the same way as above.

3.4.6 Use of historical data to improve the estimate of Between Replicates error

It will be noted that in the significance tests described above, both differences between treatments and linear trend are assessed against the variation between independent replicates of the same treatment. Since the maximum practical size of a mammalian mutation assay is likely to include only five or six duplicate treatments, this Between Replicates error is likely to be poorly estimated in a single experiment. The critical values in Variance Ratio Tables allow for this, but although the frequency of false positive results may be accurate, the test will be conservative and small genuine effects will be missed. It is therefore highly desirable to use as much data as possible to improve the estimate of Between Replicates error. Procedures for combining data from repeat experiments have been outlined in Section 3.4.5 above. It may be possible, however, to use historical data from similar, as well as repeat, experiments to improve the estimate of

Between Replicates error. We would stress that we do *not* advocate the use of historical control *means* except as a form of quality control. The use of historical data to improve an estimate of error may, however, be justified in appropriate cases.

The use of weighted analysis of variance has interesting implications for the value of the Between Replicates mean square term. If empirical weighting (Section 3.4.2.2) is used, and if there is no source of variation between replicates other than the variation between plates of the same replicate, the Between Replicates mean square term will approximate to 1.0. If, as is likely for the reasons discussed in Section 3.2, the Between Replicates mean square term is greater than 1.0, this is a measure of the additional sources of variation between replicates of the same treatment. If Poisson, rather than empirical, weighting is used and the actual variation between replicate plates is greater than Poisson, the analysis of variance will be underweighted and the size of the Between Replicates mean square term will reflect this.

Thus, where the same experiment is performed on different test substances over a period of time, the Between Replicates mean square term should be similar in different experiments and should be to some extent a measure of quality control. Although the term will be poorly estimated in an individual experiment, an unusually high value in an individual experiment, or an increasing trend in values, should be cause for concern. It is similar, but does not correspond exactly, to the 'heterogeneity factor' proposed by Robinson *et al.* (Chapter 4) for fluctuation data.

Where it is expected that the Between Replicates error term will approximate to the same value in different experiments, it should be possible to combine data to improve the estimate. This will only be appropriate where the experiments are of the same type (for example, using the same cell strain, selective system and experimental protocol, and performed by the same operators in the same laboratory). When it is possible, however, it will be expected to substantially improve the precision of the analysis.

The simplest estimate of the combined mean square from different experiments is

$$\frac{\sum(\text{Between Replicates sum of squares})}{\sum(\text{Between Replicates degrees of freedom})}$$

with the degrees of freedom being the sum of the Between Replicates degrees of freedom.

When a database of similar experiments has been built up, it becomes possible to test the acceptability of the estimate of the Between Replicates mean square for each new experiment. If the new estimate is lower than the existing combined estimate, it should be included. If the new estimate is higher, the ratio of the new and present combined estimates should be determined as a variancc ratio, with n_1 degrees of freedom for the new estimate and n_2 degrees of freedom for the present combincd estimate. If this ratio is significant at, say, the 1% level, it is likely that the new experiment shows excessive variability, possibly the result of technical error. Such an estimate should not be combined with the existing estimate, and the experiment should be discarded in whole or part.

3.4.7 Treatment of outliers

A common problem arising after an experiment has been performed and the data from it collated is that values show up out of accord with the rest. Such outlying values can seriously influence the analysis and hence the inferences drawn. Removal of such outliers prior to analysis is a contentious issue. After the data have been collected they should be reviewed for outliers. If outliers are identified the experimental records should be searched for causes for these values being corrupted. If cause is found, removal from analysis is reasonable. In the absence of cause the analysis should be performed twice, once with the outliers included, and once with them excluded, and interpretation of the results should incorporate a comparison of the two analyses.

3.4.8 Use of statistical packages

The calculations described in Section 3.4 can all be performed by hand calculator. Weighted analysis of variance and linear regression are, however, standard statistical techniques and should be available on any standard computer statistics package. For instance, with the package GLIM (Payne *et al.*, 1986), mutant frequencies are calculated and entered, mutant frequency is set as the dependent variable, a normal distribution for error is assumed, with an identity link between mutant frequency and variates such as dose, and weighted fits of alternative models are performed.

3.5 DISCUSSION

In order to be able to extract the maximum amount of information unambiguously from experimental data the design of the experiments is crucial. Two points are important to the design of mammalian cell gene mutation assays. First, sufficient cells should be plated, and sufficient plates used such that at least one plate in the series has a non-zero count. A complete series of zero counts is effectively giving no information of use concerning mutagenicity. It is recognised that for some of the assays (e.g. V79 and CHO) there might be problems in plating an adequate number of cells. It could be considered a good reason for preferring the L5178Y mouse lymphoma assay or the suspension culture variant of the CHO assay that they do not have this problem. Secondly, investigation of available data (and our experience) lead us to feel that the 'pure experimental error' against which everything is compared for significance should include a between cultures component. This is only brought in if the assay is done with more than one culture at each treatment/dose. These should be genuine replicate cultures set up independently at the start of the experiment. Failure to do this, using only one culture per treatment/dose and using only the variation between plate counts to estimate 'pure experimental error', might be the explanation for some of the results that have led to criticisms of this class of assay, because it is likely that this error will have been seriously underestimated. It is strongly recommended that at least two genuinely independent cultures be used throughout an experiment.

The paper by Leong *et al.* (1985) attempts to model the variance of the mutant frequency by compounding theoretical variances (i.e. assuming specific distributions) for different stages of the experiment. Although it does not mention the need for genuine replicate cultures its recognition of three inputs to this variance (from the initial set up of the culture, from the growth/re-sampling process, and from the plating process) endorses the concept that plate count variation is an underestimate of 'pure experimental error'. In fact it assumes that the variation between cultures is that of a Poisson distribution. In actuality the variation between cultures is likely to be greater than that of a Poisson distribution.

The data set available to the working party included a number of situations where the assay of a substance had been repeated. This made possible investigation of whether differences between experiments done at different times existed. Some differences of this sort were found and therefore it is recommended that an experiment should be repeated at least once. As an alternative to the simplified approach in Section 3.4.5, results from two (or more) experiments can be combined and analysed as one entity taking 'experiment' as a factor in the analysis. If professional statistical expertise is available this is the best approach. However, if such expertise is not available then the (relatively) simple methods described in Section 3.4.5 can be used.

Ideally, both experiments should show positive and negative controls within reasonable limits, results at each dose level should be consistent with each other, and the linear trend should be consistent between experiments. Then if the mutant frequency at one or more doses of test substance is significantly greater than that of the negative control in the combined test and if there is a significant dose relationship, the test substance should be considered positive with regard to the assay used. If no treatment shows a significant increase in mutant frequency over the negative control and there is no significant dose relationship, the test substance should be considered a non-mutagen. When there is disagreement between experiments about the significance of the effect of a dose, or about the dose relationship, or where controls are outside reasonable limits, a further experiment should be done. This should also be done where a test is of borderline significance, or where there is an effect at only one dose level, or a trend with no individual dose level being significant.

Another feature of assay design requiring comment is the number of doses of positive controls and test substance. Given the role of the positive controls as a 'quality check', one dose is felt sufficient. For the test substance a minimum of three doses is recommended. The more doses of test substance used the more information can be extracted about the relationship between mutant frequency and dose.

The purpose of the statistical tests described here is to decide whether a test substance produces a significant increase in mutant frequency, or whether an observed increase is likely to have arisen by chance. Other factors must be taken into account in interpreting a result, not least a consideration as to why an increase in mutant frequency may have occurred or why it may not have occurred. The calculations given in Section 3.4 produce results felt to match a mutagenic effect for the test substance in that there are significant 'Treatment' and 'Lin. rel.' effects in the analysis of variance table together with an appropriate pattern of means and a highly significant non-zero positive slope. Comment about attaining a threshold value is not relevant to these artificial data.

In Section 3.4 it was pointed out that the negative control data can either be treated as a separate treatment or combined with the test substance data for the purposes of calculating the linear dose relationship. The argument in favour of combination is that negative control is zero dose of the test substance and as such provides information about its relationship with dose. The counter argument is that a straight line is only an approximation to the actual dose relationship and incorporating the negative control data into the calculations worsens the approximation and (usually) will bias the slope downwards (toward zero) which may reduce the discriminating power in consequence. It should be noted, however, that for the model data analysed in Section 3.4, although inclusion of the negative control does indeed reduce the slope and increases the non-linear error component, it does not in fact reduce the significance of the linear component. It may, however, make the linear relationship calculations more sensitive to the location of the effective dose range, i.e. two mutagenic test substances with the same *MF* values for treatments and negative control will give different linear relationships if the dose range of one is from 10 to 30 (say) and that of the other from 110 to 130.

3.6 SUMMARY OF RECOMMENDATIONS

(a) Treat sufficient cells.
(b) Use 'sufficient' plates and plate 'sufficient' cells.
(c) Use at least two genuinely independent cultures throughout.
(d) Analyse the Adjusted (mean) Mutant Frequencies at each treatment/dose, not the individual plate counts. Only use the latter to calculate weights.
(e) Analyse by weighted analysis of variance (regression) (Section 3.4.2) using either Poisson-derived or empirically-derived weights.
(f) Test mutant frequency at each treatment dose for a significant increase over the negative control.
(g) Test for a significant linear relationship between increasing mutant frequency and increasing dose.
(h) Repeat all experiments.
(i) Compare and combine results from repeated experiments.

3.7 ACKNOWLEDGEMENTS

In addition to the data sets provided by members of the working group we are indebted to M. O'Donovan (The Boots Company plc, The Priory, Thurgarton, Nottingham NG14 7GX) and M.G. Clare (Shell Research Ltd, Sittingbourne Research Centre, Sittingbourne, Kent ME9 8AG) for the use of their Chinese hamster data.

3.8 REFERENCES

Amacher, D.E., Paillet, S.C., Turner, G.N., Ray, V.A. & Salsburg, D.S. (1980). Point mutations at the thymidine kinase locus in L5178Y mouse lymphoma cells. 2. Test validation and interpretation. *Mutation Research*, **72**, 447–74.
Amacher, D.E. & Salsburg, D.S. (1981). Letter to the Editor. *Mutation Research*, **89**, 245–9.
Box, G.E.P. & Cox, D.R. (1964). An analysis of transformations (with discussion). *Journal of the Royal Statistical Society, B*, **26**, 211–46.
Boyd, M.N. (1982) Examples of testing against ordered alternatives in the analysis of mutagenicity data. *Mutation Research*, **97**, 147–53.
Carver, J.H., Adair, G.M. & Wandres, D.L. (1980), Mutagenicity testing in

mammalian cells. II. Validation of multiple drug-resistance markers having practical application for screening potential mutagens. *Mutation Research*, **72**, 207–30.

Clive, D. & Hajian, G. (1981). Letter to the Editor. *Mutation Research*, **89**, 250–3.

Clive, D., Hajian, G. & Moore, M.M. (1981). Letter to the Editor. *Mutation Research*, **89**, 241–4.

Clive, D., Johnson, J.F.S., Spector, A.G., Batson, A.G. & Brown, M.M. (1979). Validation and characterization of the L5178Y/TK mouse lymphoma mutagen assay system. *Mutation Research*, **59**, 61–108.

Cole, J. & Arlett, C.F. (1984). The detection of gene mutations in cultured mammalian cells. *In: Mutagenicity Testing – A Practical Approach*, ed. S. Venitt and J.M. Parry, IRL Press, Oxford, pp. 233–73.

Cole, J., Fox, M., Garner, C., McGregor, D. & Thacker, J. (1983). Gene mutation assays in cultured mammalian cells. In: *UKEMS Sub-committee on Guidelines for Mutagenicity Testing*. Report. Part 1: Basic test battery; ed. B.J. Dean. United Kingdom Environmental Mutagen Society, Swansea, pp.65–102.

Dunnett, C.W. (1955). A multiple comparison procedure for comparing several treatments with a control. *American Statistical Society Journal*, 1096–120.

Furth, E.E., Thilly, W.G., Penman, B.W., Liber, H.L. & Rand, W.M. (1981). Quantitative assay for mutation in diploid, human lymphoblasts using microtiter plates. *Analytical Biochemistry*, **110**, 1–8.

Hsie, A.W., Brimer, P.A., O'Neill, J.P., Epler, J.L., Guerin, M.R. & Hsie, M.H. (1980). Mutagenicity of alkaline constituents of a coal-liquefied crude oil in mammalian cells. *Mutation Research*, **78**, 79–84.

Hsie, A.W., Casciano, D.A., Crouch, D.B., Krahn, D.F., O'Neill, J.P. & Whitfield, B.L. (1981). The use of Chinese hamster ovary cells to quantify specific locus mutation and to determine mutagenicity of chemicals. A report of the Gene-tox program. *Mutation Research*, **86**, 193–214.

Irr, J.D. & Snee, R.D. (1982). A statistical method for analysis of mouse lymphoma L5178Y cell TK locus forward mutation assay. *Mutation Research*, **97**, 371–92.

Lea, D.E. & Coulson, C.A. (1949). The distribution of numbers of mutants in bacterial populations. *Journal of Genetics*, **49**, 264–85.

Lee, Y.J., and Caspary, W.J. (1983). Mathematical model of L5178Y mouse lymphoma forward mutation assay. *Mutation Research*, **113**, 417–30.

Leong, P.-M., Thilly, W.G. & Morgenthaler, S. (1985). Variance estimation in single-cell mutation assays: comparison to experimental observations in human lymphoblasts at 4 gene loci. *Mutation Research*, **150**, 403–10.

Moore-Brown, M.M., Clive, D., Howard, B.E., Baston, A.G. and Johnson, K.O. (1981). The utilisation of trifluorothymidine (TFT) to select for thymidine kinase-deficient (TK −/−) mutants from L5178Y/TK+/− mouse lymphoma cells. *Mutation Research*, **85**, 363–78.

Newbold, R.F., Amos, J. & Connell, J.R. (1979). The cytotoxic, mutagenic and clastogenic effects of chromium-containing compounds on mammalian cells in culture. *Mutation Research*, **67**, 55–63.

O'Neill, J.P. & Hsie, A.W. (1979). Phenotypic expression time of mutagen-induced 6-thioguanine resistance in Chinese hamster ovary cells (CHO/HGPRT system). *Mutation Research*, **59**, 109–18.

Payne, C.D., Atkin, M., Baker, R.J., Clarke, M.R.B., Francis, B., Gilchrist, R., Green, M., Nelder, J.A., Reese, R.A., Slater, M., Swan, A.V. & White, R. (1986). The GLIM system, release 3.77. *Numerical Algorithms Group, Oxford.*

Robinson *et al.* this volume, Chapter 4.

Snee, R.D. & Irr, J.D. (1981). Design of a statistical method for the analysis of mutagenesis at the hypoxanthine-guanine phosphoribosyl transferase locus of cultured Chinese hamster ovary cells. *Mutation Research*, **85**, 77–93.

Tan, E.-L. & Hsie, A. (1981). Mutagenicity and cytotoxicity of haloethanes as studied by the CHO/HGPRT system. *Mutation Research*, **90**, 183–91.

Topping, J. (1972). *Errors of Observation and their Treatment*. 4th ed. Chapman & Hall, London, p.82.

Van Zeeland, A.A., Van Diggelen, M.C.E. & Simons, J.W.I.M. (1972). The role of metabolic co-operation in selection of hypoxanthine-guanine-phosphoribosyl-transferase (HG–PRT)-deficient mutants from diploid mammalian cell strains. *Mutation Research*. **14**. 355–63.

4

Statistical evaluation of bacterial/mammalian fluctuation tests

W.D. ROBINSON
M.H.L. GREEN, (Group leader)
J. COLE M.J.R. HEALY
R.C. GARNER D. GATEHOUSE

4.1 INTRODUCTION

4.1.1 Background

The fluctuation test was originally devised by Luria and Delbruck (1943) in order to demonstrate that bacterial variants arose by random mutation rather than by adaptation to a selective agent. They argued that if variants arose by adaptation, the numbers of mutants found in a series of small independent replicate cultures should be rather uniform whereas, if variants arose by mutation, there should be a wide variation. They successfully confirmed this second prediction. Their experiment also allowed an estimate of mutant frequency, most conveniently by equating the fraction of replicate cultures containing no mutants to the zero term of the Poisson distribution

$$m = -\ln(P_0)$$

where P_0 = fraction of cultures containing no mutants and m = average number of mutants per culture.

The fluctuation test has been adapted to measure induced mutation in bacteria (for review, see Hubbard *et al.*, 1984). In such experiments, a number of sets of replicate cultures are prepared. Each set contains cells treated with one dose of test agent. The experiments are designed so that a proportion of the replicate cultures will contain no mutants and the zero term of the Poisson distribution can be used to estimate mutant frequency. Should the agent under test be mutagenic, the proportion of cultures containing mutants will increase.

The fluctuation test has also been modified for use in mammalian cells (see Cole *et al.*, 1983*a*) and has proved particularly convenient for the assay of mutation in mammalian cells such as L5178Y or TK6 which are grown in suspension culture (Furth *et al.*, 1981).

Fluctuation tests are now routinely performed in multiwell trays (plates) and in this situation each well is equivalent to an individual small replicate culture in the original Luria and Delbruck experiment. One or more trays will be used for each treatment.

4.1.2 Mammalian fluctuation tests

A flow diagram for a typical mammalian fluctuation test using L5178Y cells is given in Fig. 4.1. Three or four days before treatment, bulk cultures are established in flasks in suspension culture medium at $1–5 \times 10^3$ cells/ml (using cell cultures purged of pre-existing mutants, or subcultured from low cell density using a routine known by the laboratory to result in a reasonably stable spontaneous mutant frequency) to provide a large population of exponentially growing cells for treatment on Day 0. The bulk cultures are pooled, counted and adjusted to a standard cell density for treatment of 5×10^5 per ml. This pooled culture is then divided into single or duplicate cultures of 20–40 ml per culture (i.e. $1–2 \times 10^7$ cells). When predicted kill following treatment is high (e.g. 50–90%) large numbers of cells may be treated and maintained in suspension culture for the expression time. The cultures are treated with test compound, and incubated. At the end of the treatment period, the cultures are centrifuged, washed to remove the test compound, resuspended in fresh medium and counted using a haemocytometer or Coulter counter.

Each culture is adjusted to a cell density of 2×10^5 per ml by adding fresh medium and at this stage no cells are discarded. An aliquot is removed from each culture, to be diluted and plated to determine survival immediately after treatment. A multichannel or microdoser pipette is used to deliver 0.2 ml of cell suspension at about 10 cells per ml into each well of one or more 96 well microtitre trays (plates). With duplicate treatments, the two cultures are kept entirely separate throughout the experiment.

The cultures are incubated at 37 °C, counted every 24 h and diluted to maintain exponential growth and to allow time for newly induced mutants to be expressed. Each day, a minimum number of cells are subcultured, commonly 1×10^7 cells at 2×10^5 cells per ml in suspension culture. When larger numbers of cells are initially treated (to allow for toxicity), more than 10^7 cells may be maintained in each culture for the first 96 h expression time, or until growth rate is normal.

When sufficient time has elapsed for induced mutants to be fully expressed (see Table 4.1), the cells are plated at high density in selective

Fig. 4.1. Flow diagram for a typical L5178Y mutation assay. Procedures for TK6 would be similar, but numbers would be varied.

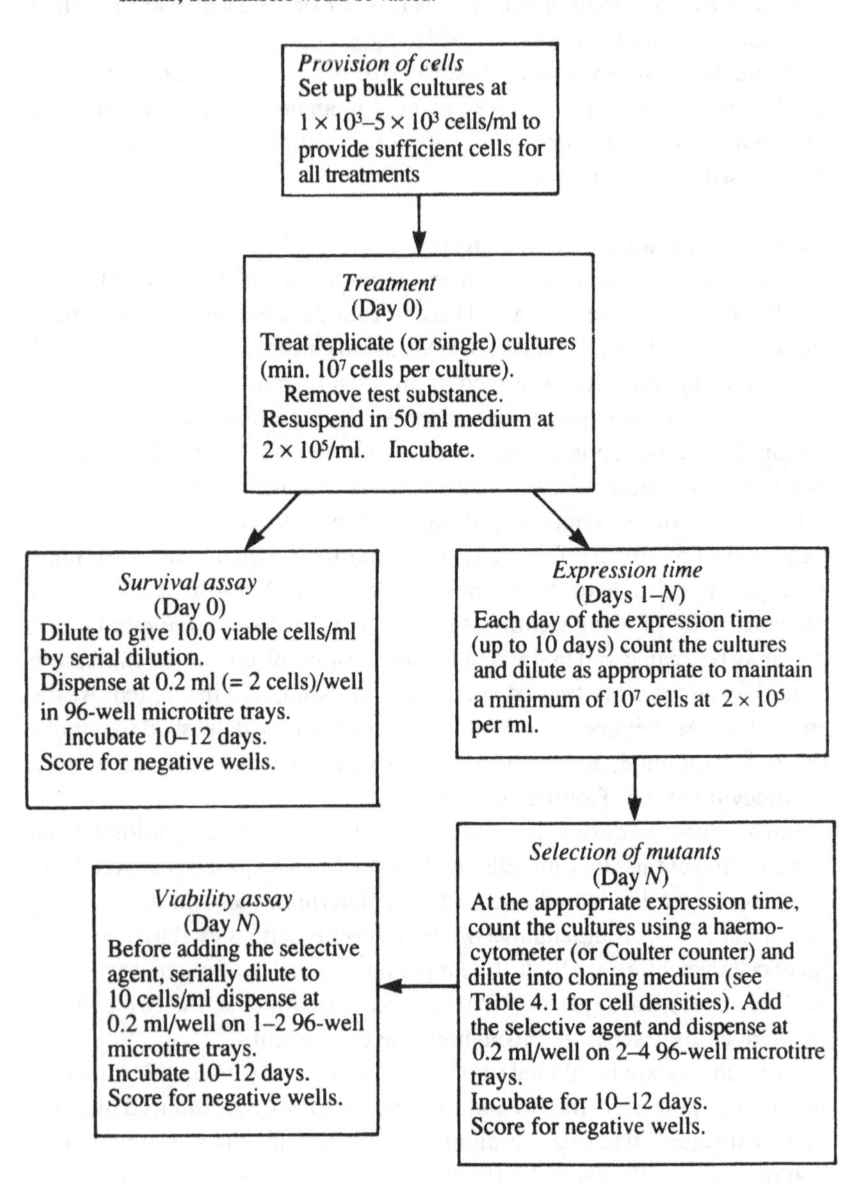

Table 4.1. *Commonly used selective systems for mammalian fluctuation assays*

Selective agent (final concentration)	Historical spontaneous mutant frequency (95% upper confidence limit)	Expression time	Max. number of cells per well	Number of trays per point	Typical % +ve control wells
Ouabain[a](10^{-3} M)	1.8×10^{-6}(4.3)	2–3 days	2×10^4	2–4	0.7%
6TG[a](15 or 30 μg/ml)	2.4×10^{-6}(6.8)	7–10 days	1×10^4	2–4	0.5%
TFT[b](2 or 4μg/ml)	11.8×10^{-5}(23.6)	2–3 days	2×10^3	2	4.5%

Notes:
[a] values for L5178Y wild-type cells for Ouabain and for 6-thioguanine (6TG) resistance.
[b] values for L5178Y TK +/− cells for Trifluorothymidine (TFT) resistance.
Source: data from J. Cole.

medium to determine mutant frequency, and at low density in non-selective medium to determine viability. At the appropriate expression time, the cells are counted (using a haemocytometer or Coulter counter), and diluted into cloning medium. Maximum satisfactory cell densities are given in Table 4.1. The selective agent is added and the cells are distributed at 0.2 ml per well into two to four 96-well microtitre trays using a multichannel or microdoser pipette. Before the addition of the selective agent, an aliquot is removed, to be diluted into non-selective cloning medium at ~ 10 cells per ml, to be plated on one or two 96-well trays at 0.2 ml per well to estimate viability of each culture. The trays are incubated for 10–12 days and are scored for negative wells (i.e. wells in which no growth has occurred) using a low-power dissecting microscope.

4.1.3 Bacterial fluctuation tests

Bacterial fluctuation tests are considerably simpler in design than those using mammalian cells and a typical flow diagram is given in Fig. 4.2. Viability is not normally assayed and the bacteria are simply mixed with the test agent (with or without activation) and dispensed in trays in an appropriate selective medium. After a period of incubation (normally 2–3 days) wells are scored visually, usually with the aid of an indicator dye, for the presence or absence of mutants (Hubbard *et al.*, 1984).

4.1.4 Published methods of statistical analysis

Green *et al.* (1976) set out the result of a bacterial fluctuation test in the form of a 2 × 2 table containing the numbers of wells in the categories positive and negative for growth, and compared treated vs control, using a standard χ^2 to test for significance. Gilbert (1980) argued that a one-sided test for an increase in mutant frequency was appropriate and used the signed square root of χ^2 (the square root with a + or − sign, depending on the direction of the effect) as a normal deviate. Collings *et al.* (1981) similarly recommended a one-sided test and reported on the statistical considerations of the fluctuation test in detail. Venitt (1982) transformed the proportion of negative wells to estimate the average number of mutants per well, assuming a Poisson distribution. He then found the slope of the dose-response curve in order to test for a mutagenic effect, although apparently without allowing for the varying precision of the estimates.

With a mammalian test, viability is determined in parallel with mutant frequency (Box 4.1, 4.2). Furth *et al.* (1981) calculated the log mutant frequency for each dose and its standard deviation based on the binomial distribution of the fluctuation test data. Mutant frequencies outside the 95% confidence limit of the control were deemed significant, provided that

Fig. 4.2. Flow diagram for a typical bacterial fluctuation test.

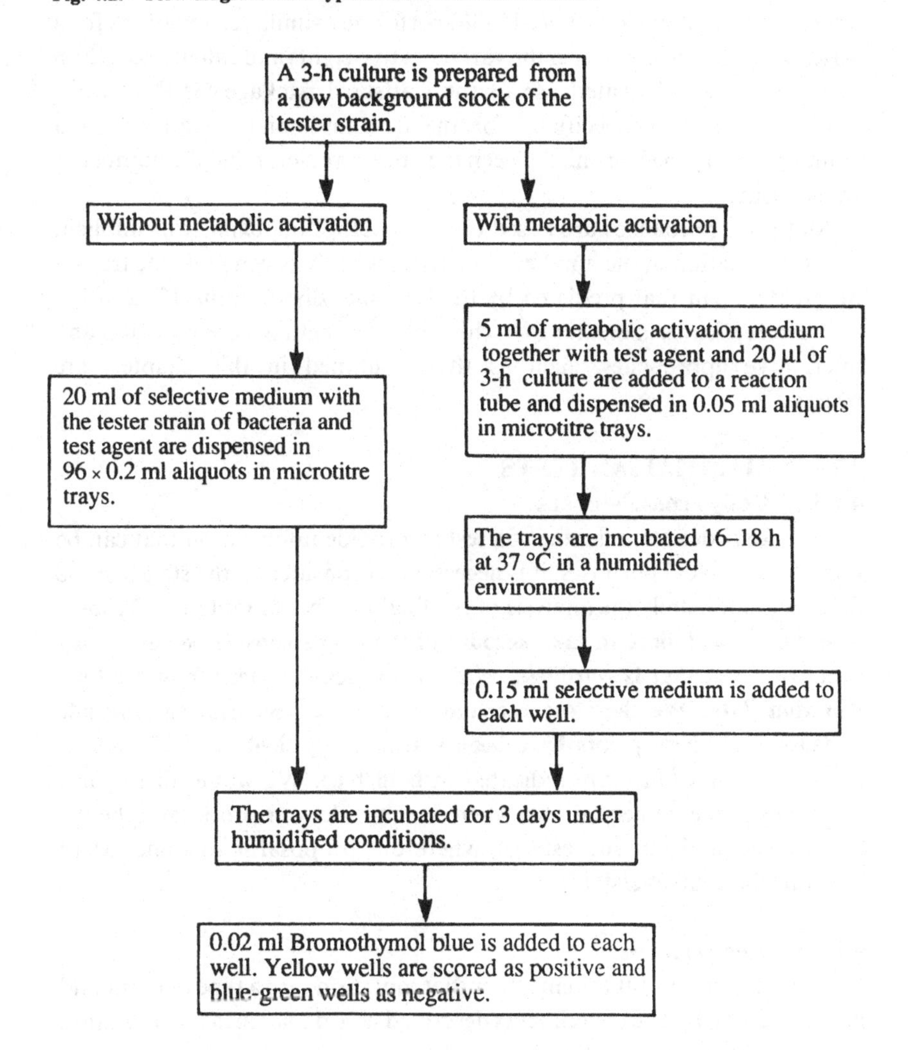

these were also outside the confidence limit of the historical control. Thomas (quoted in Cole *et al.*, 1983*a*) performed similar calculations for a series of doses, but converted the variances to weights and calculated a form of χ^2. He also advocated use of the statistical package GLIM (Royal Statistical Society), specifying for the data binomial variation and a complementary log/log link between the linear predictor and the number of wells positive.

All these approaches are based on the assumption, explicit or implicit, that the variation in the number of negative wells between replicate trays is no greater than that predicted by the binomial distribution. If variation between replicates is greater than this, then the methods are not valid and alternative approaches, such as those outlined in this chapter, are necessary.

4.2 MAMMALIAN TESTS

4.2.1 Design considerations

In order for a fluctuation test to provide information that can be used for statistical purposes, it is necessary to consider all the stages of the protocol, not merely the analysis of the final number of mutants. **We must emphasise that failure to take account of the points considered below may result in a test that is worthless, without this being evident from the final mutation data. We therefore recommend that test reports must include evidence that these points have been satisfactorily dealt with.** Points of particular importance are indicated in **bold type**. We would distinguish between our recommendations, where we consider our proposal to be the correct one, and our suggestions, where our proposal is to some extent arbitrary or controversial.

4.2.1.1 *Stock cultures*

It is important to remember that mutations are a type of event and mutants are cells. The experiments described here do *not* measure mutation rate (the rate at which mutations occur) but measure mutant frequency (the frequency of mutant cells in the population). The mutants scored may have been present in the culture before the start of the experiment as a result of mutation events in the past, they may be mutants arising from mutation events during the course of the experiment, or they may be descendants of mutants arising from mutation events during the experiment. In any large population of cells (bacterial or mammalian) the frequency of pre-existing mutants will be substantially higher than the frequency of spontaneous mutants arising during a given period, and the greater the ratio of pre-existing to newly arising spontaneous mutants the less sensitive will be the test. **Hence stock cultures must be maintained with as low a frequency of pre-existing mutants as possible, in order for the test to be sensitive** (Cole & Arlett, 1984). The level of pre-existing mutants will of course be higher in systems with a higher spontaneous mutation rate; it is the ratio of pre-existing to newly arising mutants which must be kept as low as possible.

Although it is important to check the negative (and positive) control values against historical data, a statistical test would be difficult and probably inappropriate. Firstly, the distribution of numbers of mutants in starting cultures is complex (Lea & Coulson, 1949) and unsuitable for a simple statistical test. Secondly, the point at issue is not simply whether an excessive number of pre-existing mutants arose by chance, but whether the excess will reduce unacceptably the sensitivity of the assay. For these reasons, we suggest simply that if the negative control mutant frequency is more than three times the historical control mean, it might be prudent to discard the experiment.

It is regrettably common practice to use the ratio of induced to spontaneous *mutant frequency* to evaluate mutagenicity data, either using a doubling as evidence of mutagenicity, or using the ratio as an estimate of mutagenic potency. Since the original point of the fluctuation test was to demonstrate the high variability of spontaneous mutant frequency, the ratio of induced mutant frequency to a highly variable spontaneous mutant frequency cannot be used as a quantitative measure of induced mutation rate. If a quantitative estimate of *mutation rate* is required, a special design of experiment must be employed (Cole *et al.*, 1976).

4.2.1.2 *Repetition of experiments*

We recommend that all results must be confirmed in an independent experiment (Cole *et al.*, 1983b). Should the two disagree, a third must be performed. A single experiment is not adequate evidence of either mutagenicity or non-mutagenicity. We describe below methods for testing separate experiments for consistency, and for combining results from separate experiments.

4.2.1.3 *Number of doses*

Since mammalian mutagenicity assays are extremely laborious, we strongly advise that an initial experiment to estimate toxicity should be performed before undertaking a full mammalian fluctuation test. **We recommend that the actual mutation test should try to estimate mutant frequency at a dose giving 10–30% survival, a dose giving 30–70% survival and a dose giving greater than 70% survival. Hence in a full experiment, provided that toxicity has been determined in advance, no more than three–**

five doses plus a negative and a positive control should be required. If toxicity increases linearly with dose, which is often not the case, the highest dose will need to be 10–20 times larger than the lowest, in order to achieve these figures. Doses in the ratio of 0:1:4:16, or 0:1:2:4:8:16 might be appropriate.

4.2.1.4 *Number of cells to be treated and subcultured*

Thilly *et al.* (1980) recommend that sufficient cells be treated so that at least 100 mutants on average should survive treatment. Assuming a Poisson distribution, this figure should in 95% of cases lead to observed counts between 80 and 120, which is considered to be acceptable variation for most analyses. With systems such as ouabain resistance or thioguanine resistance, the low spontaneous mutant frequency makes it impractical to maintain this number in an experiment of acceptable size. If only ten mutants were to be aimed for, then the equivalent range of observed counts would be about 4 to 16.

If, at each subculturing, the population is diluted then the variance of the population of mutants available for plating will increase with the number of subculturings (Leong *et al.*, 1985). **Subculturing introduces variation that cannot be estimated from a single culture. Therefore we recommend that treatment and control subcultures must be replicated, otherwise it will not be possible to carry out a valid test of significance. This will be the case, whether mutant frequency is measured in a fluctuation test or by a plating assay in which colonies are scored.**

If it is expected that the treatment will be toxic, so that only *S*% of the cells survive, then the number of cells carried at the first subculture should be multiplied by 100/*S*, and additional cells should be carried at each subsequent subculture, until non-viable cells are diluted out, in order to maintain approximately the same statistical precision. We suggest that, after allowing for toxicity, twice as many cells be carried and also twice as many trays be plated per culture for the untreated controls as for each treatment. (Carrying more cells will improve precision, but unless more trays are plated, this improved precision will not be reflected with the form of statistical test which we advocate.) Having twice the replication for the controls as for each treatment is the optimum allocation of resources when there are four treatments to compare with the control. Our suggestion is similar in effect, but in this instance using larger control samples has some slight advantages over using more controls of the same size.

4.2.1.5 *Number of test cultures*

We recommend that controls should be subcultured in duplicate, and that all treatments should be carried out and subcultured in duplicate.

Table 4.2. *Recommended size of mammalian fluctuation assays*

Treatment	Number of cultures per treatment level	Cells/culture treated and carried at first subculture	Trays per replicate culture			
			Survival	oua^{res}	6TG^{res}	TFT^{res}
Control	2	2×10^7	2	8	8	4
100–70% survival	2	1×10^7 (1.5×10^7)	1	4	4	2
70–30% survival	2	2×10^7 (1.5×10^7–3×10^7)	1	4	4	2
30–10% survival	2	5×10^7 (3×10^7–10×10^7)	1	4	4	2

Carrying a larger number of replicate treatments would be desirable but impractical. We believe that duplicate treatments at a small number of carefully selected doses is likely to be the most efficient use of resources.

4.2.1.6 ***Number of cells per well and number of trays***

In estimating mutant frequency from a multiwell tray, optimum sensitivity occurs when 50–80% of the wells are positive (Fisher, 1954; Green *et al.*, 1976; Collings *et al.*, 1981). It is often not possible to achieve this situation, especially when assaying for mutants in the presence of selective agent, where relatively low cell numbers must be plated in order to distinguish mutants clearly. Forty multiwell trays, for instance, might be required in order to assay a population of cells likely to contain 100 spontaneous 6-thioguanine resistant mutants. Although increasing the number of trays involves a relatively small amount of extra labour, there clearly are limitations, and 4–8 trays per replicate is probably the practical limit.

4.2.1.7 ***Overall size of recommended test***

Table 4.2 summarises the size of experiment that we recommend. An adequate test (Cole *et al.*, 1983*b*) should include an independent repeat experiment, treatment at 3–5 carefully selected doses plus untreated and positive controls, with all treatments and controls in duplicate. At least 1×10^7 cells should be carried at each subculture. One 96-well tray plated with 1.5–2 viable cells per well should be sufficient to measure survival. Since it is not normally possible to obtain 1.5–2 mutant cells per well, we suggest that, with a more mutable locus such as trifluorothymidine resistance, at least two 96-well trays for each replicate culture be plated to

Box 4.1. *Simulated data for use in the mammalian fluctuation worked examples. The total wells are not necessarily exact multiples of 96, because one or more wells may be lost through contamination or be unscorable*

		Survival			Mutation	
Dose replicate (i	Wells neg. y_s	Total wells n_s	Cells/ well f_s	Wells neg. y_m	Total wells n_m	Cells/ well f_m)
0 1	32	192	2.0	717	768	4000
2	36	192	2.0	724	765	4000
1 1	14	96	2.0	362	384	4000
2	16	95	2.0	351	378	4000
2 1	21	90	2.0	347	384	4000
2	34	96	2.0	358	384	4000
4 1	8	96	2.0	359	384	4000
2	34	96	2.0	359	384	4000
8 1	30	95	2.0	341	384	4000
2	24	96	2.0	328	384	4000
16 1	25	96	2.0	299	380	4000
2	24	96	2.0	311	384	4000

Notes:
Simulated data shown for replicate cultures 1 and 2.
2 pooled trays/culture for negative control for survival.
1 pooled tray/ culture for treatment cultures for survival.
8 pooled trays/culture for negative control for mutation.
4 pooled trays/culture for treatment cultures for mutation.
f_s = 2.0 cells/well estimated from 10 cells/ml, 0.2 ml/well.
f_s = 4000 cells/well estimated from 2×10^4 cells/ml, 0.2 ml/well.

measure mutant frequency. With a less mutable system such as 6-thioguanine resistance, at least four trays should be used. These numbers may still be less than ideal. As indicated in Section 4.2.1.4 above, twice as many cells should be carried, and twice as many trays plated for the negative control series.

Box 4.1 shows 'typical' data. Boxes 4.2–4.9 show worked examples of the recommended analyses using these data. These are described below in Sections 4.2.2–4. Although the calculations are given here to three places of decimals, in practice the full accuracy of the calculator or computer should be used, otherwise substantial rounding errors may be introduced.

4.2.2 Heterogeneity factor

The data from all the methods we consider take the form of *counted proportions*, r negative wells (say) out of a total of n giving a proportion of

$p = r/n$. Under the best possible circumstances, the variability of such a proportion can be obtained from the binomial distribution as a variance of $p(1-p)/n$. We have found that the variability of the proportions negative on two or more plates prepared from a single culture usually agrees quite well with this theoretical value (see Section 4.2.4.3). However, when trays are prepared from different cultures which have been grown under similar conditions, extra sources of variability are introduced by the successive dilutions and other manipulations, especially during subculturing (Leong *et al.*, 1985). As a result, the proportions negative from such trays have a larger variance than the theoretical binomial value. We have calculated the observed variances from several experiments, and we find that the ratio of these variances to the theoretical binomial variances is roughly constant. We refer to this ratio as the *heterogeneity factor*, a term taken from its comparable use in biological assay (Finney, 1978). In our recommended calculations, all the variances and standard errors which form the basis for significance tests and confidence limits are obtained by multiplying a binomial variance by an appropriate heterogeneity factor.

It is theoretically possible to estimate the heterogeneity factor from the results of a single experiment. However, such an estimate will inevitably be very imprecise, and using it will degrade the precision of the results. Instead, we recommend that each laboratory should use factors based on their own previous experience. When each experiment is done, a new value for the factor is calculated and compared with the 'historical' value. If the new value is exceptionally large, the experimental results are unacceptably variable and must be discarded. Otherwise, the historical value is updated and used in the analysis. We give detailed methods for these calculations in subsequent sections.

The heterogeneity factor is likely to vary to some extent according to the experimental conditions. We have found values of 3.0 for survival trays and 1.8 for mutation trays. However, we do not suggest that these values be accepted by other workers as other than starting values upon which local values can be based (see Sections 4.2.2.3, 4.2.4.1)*.

* Examination of the wider range of data generated in the UKEMS third trial, suggests that use of higher starting values may be advisable. Moreover, if there is a heterogeneity factor but it is not proportional to the underlying binomial distribution, then it may be affected by such factors as the number of trays per replicate or the number of wells positive per tray. Such a situation might also arise if routine estimates of the number of cells plated per well were inaccurate and the same estimate were used for plating both viability and mutation trays. We suggest it may be advisable to confine re-estimates of the heterogeneity factor to similar experiments, and to maintain different estimates for different designs of experiment.

Table 4.3. *Comparison with current heterogeneity factor, largest acceptable ratios (one-sided* F*-test assuming current factor is known)*

Degrees of freedom	1	2	3	4	5	6		
Cultures (0.1%)	10.8	6.9	5.4	4.6	4.1	3.7		
Degrees of freedom	1	2	3	4	5	6	7	8
Experiment (1%)	6.6	4.6	3.8	3.3	3.0	2.8	2.6	2.5
Degrees of freedom	9	10–11	12	13–14	15–17	18–20		
Experiment (1%)	2.4	2.3	2.2	2.1	2.0	1.9		

Source: from Pearson & Hartley (1976).

4.2.2.1 *Determination of the heterogeneity factor for a single experiment*

The case where there are two replicate cultures is the simplest to describe and is the one we consider first. Box 4.2 gives a worked example, using the zero dose data from Box 4.1. We shall assume in this section, that if several trays were used per culture then they were sufficiently consistent to allow us to sum over the trays, the numbers of negative wells and the total wells. Later we indicate how to test this assumption.

For the control and each treatment we have:

	Wells negative	Total wells	Proportion
Culture 1	y_1	n_1	$p_1 = y_1/n_1$
Culture 2	y_2	n_2	$p_2 = y_2/n_2$
Sum	$Y = y_1 + y_2$	$N = n_1 + n_2$	$P = Y/N$

The best estimate of the proportion of negative wells is P, which should therefore be used in the calculation of variances based on the assumption of a binomial distribution and of equality of cultures.

Compute the difference in proportions

$$d = p_1 - p_2$$

and its variance

$$V = P(1 - P)(1/n_1 + 1/n_2)$$

Then

$$h = d^2/V$$

would be expected to exceed 10.8 times the current heterogeneity factor (H) only rarely. (10.8 is the one-sided 0.1% level of the F-distribution with 1 and infinite degrees of freedom assuming the current factor to be known. The level has been chosen to compensate for the large number of comparisons

Box 4.2. *Worked example of test for consistency of duplicate cultures at each dose level*

Existing survival heterogeneity factor = 1.8 (H_s)
Existing mutation heterogeneity factor = 3.0 (H_m)

	Survival data, dose 0			Mutation data, dose 0		
Culture	Wells neg.	Total wells	Proportion	Wells neg.	Total wells	Proportion
	y	n	$p = y/n$			
1	32	192	0.1667	717	768	0.9336
2	36	192	0.1875	724	765	0.9464
	Y	N	$P = Y/N$			
	68	384	0.1771	1441	1533	0.9400

The difference in proportions is computed

$d = p_1 - p_2$
$= 0.1667 - 0.1875$
$= -0.0208$

$= 0.9336 - 0.9464$
$= -0.0128$

The variance, V of d is

$V = P(1-P)(1/n_1 + 1/n_2)$
$= 0.1771(1-0.1771)(1/192 + 1/192)$
$= 0.001518$

$= 0.9400(1-0.9400)(1/768 + 1/765)$
$= 0.0001472$

The square of the difference in proportions is divided by the variance. Values 10.8-fold higher than the current heterogeneity factor are excluded.

$h = d^2/V$
$= (-0.0208)^2/0.001518$
$= 0.285$

$= (-0.0128)^2/.0001472$
$= 1.113$

The process is repeated for each dose level

	H_s 10.8 = 19.44	H_m 10.8 = 32.4
	h_s	h_m
Dose 0	0.285	1.113
Dose 1	0.184	0.629
Dose 2	3.260	2.101
Dose 4	20.602[a]	0.000
Dose 8	1.019	1.953
Dose 16	0.027	0.633

[a] The difference between survival duplicates at dose level 4 is more than 10.8-fold higher than the current heterogeneity factor and this dose level is excluded from further consideration.

made in any experiment. It appears in Table 4.3, 'Cultures' row.) We recommend that pairs of cultures yielding such high values of h should be excluded from analysis. This would avoid undue influence being exerted by what are likely to be the outcomes of technical errors. The mean of the hs can then be calculated for the (possibly reduced) set of pairs of cultures obtained under the conditions of interest. The mean value of h is the heterogeneity factor for these conditions. The conditions might specify type of tray, expression time, metabolic activation, and screening system. We have found the only conditions we have to distinguish are type of tray, that is whether it is for survival or mutation. We denote their corresponding heterogeneity factors by H_s and H_m. Our current estimates are 1.8 and 3.0 respectively, but it must be emphasised that each laboratory should determine and update its own values.

4.2.2.2 *Consistency check*

The number of pairs of cultures included in the calculation of the heterogeneity factor determines its precision and reliability. Usually the number will be small for a single experiment, making the estimate inevitably unreliable. It should be combined with the current factor for testing purposes. However, first the two must be checked for consistency. The 'Experiment' rows of Table 4.3 show for various numbers of pairs (equivalent to degrees of freedom in the table) the largest acceptable ratio of the new estimate to the current factor. These rows are based on the one-sided 1% critical values from the F-distribution. Again this is an arbitrary level, chosen to allow for change, but to exclude technical error.

If the ratio is exceeded the whole experiment should be discarded and repeated afresh. It would be inadvisable to use the observed factor in these circumstances for testing because it would be unrealistically large and poorly estimated being derived from (relatively) few observations.

It could also happen that the new estimate is so much smaller than the current one that the inverted ratio, that is the ratio of the current to the new estimate, would exceed the tabulated value. Such an occurrence should not be treated specially; the observed factor, which is poorly estimated on its own, should still be combined with the current factor as described in the next section.

Worked examples of this and the following procedure are set out in Box 4.3.

4.2.2.3 *Combining new and current estimates*

Having demonstrated consistency, we should combine the new estimate with the current one. Our suggestion is to define the updated

Box 4.3. *Worked example of the test for overall consistency, and of the calculation of new heterogeneity factors*

Calculate the mean values for h from Box 4.2, for doses not excluded

$$H_{exp} = (h_1 + h_2 \ldots + h_N)/N$$

where there are N doses, including the control

$$H_{s\ exp} = (0.285 + 0.184 + 3.260 + 1.019 + 0.027)/5 = 0.955$$

$$H_{m\ exp} = (1.113 + 0.629 + 2.101 + 1.953 + 0.633)/5 = 1.286$$

Calculate the ratio of the new estimate (H_{exp}) to the current estimate (H_{old}). If it exceeds the value for that number of degrees of freedom in the 'Experiment' row of Table 4.3, the experiment should be discarded. For five pairs the critical value is 3.0.

$$\text{ratio, } H_{s\ exp}/H_{s\ old} = 0.955/1.80 = 0.531$$

$$\text{ratio, } H_{m\ exp}/H_{m\ old} = 1.286/3.00 = 0.429$$

Since these ratios are less than the critical value (3.0), calculate new estimates for the heterogeneity factors

$$H_{s\ new} = H_{s\ old}\ 19/20 + H_{s\ exp}/20 = 1.80 \times 19/20 + 0.955/20 = 1.758$$

$$H_{m\ new} = 3.0 \times 19/20 + 1.286/20 = 2.914$$

(combined) estimate as the sum of 1/20 the new estimate and of 19/20 the current factor. This weighting of 1/20 and 19/20 is an arbitrary but pragmatic compromise; it should reduce wild fluctuations in the heterogeneity factor whilst allowing changes to take place gradually. The proposed method of combining estimates gives an exponentially weighted average of the individual estimates from all experiments. This is a well-known procedure for smoothing short-term fluctuations whilst retaining long-term trends in data.

4.2.2.4 *Historical controls*

We suggest that the same procedure of weighting 1/20 of the current estimate and 19/20 of the historical estimate can be used as a convenient method of maintaining and updating historical negative and positive control data. We explain in Section 4.2.1.1 the difficulties of a statistical test for departure of negative control mutant frequencies from historical values. We therefore suggest that those experiments should be discarded where the negative control mutant frequency is more than three times the historical mean value, or where the difference between the positive and negative control mutant frequencies is less than half the historical mean value. We would stress that a drift in historical mutant frequencies may be cause for concern since it may reflect a build up of pre-existing mutants, or an alteration in properties of the tester cell line, or an alteration in the conditions of the test.

4.2.3 Recommended analyses

We recommend that two separate types of statistical test be used on the data from a single mammalian fluctuation experiment. These amount to first comparing the control log mutant frequency (*LMF*) with the *LMF* from each treatment dose, and secondly checking whether there is a linear trend in mutant frequency with treatment dose. Both tests require the use of the heterogeneity factor to obtain a modified estimate of variance. Both tests use standard statistical procedures, but it has been possible to simplify some of the calculations for this particular type of data.

4.2.3.1 *Calculation of log mutant frequency and use of the heterogeneity factor to calculate variance of log mutant frequency and statistical weight of mutant frequency*

Using the notation given in Box 4.1, the mutant frequency (*MF*) is calculated as

$$MF = \frac{-\ln(y_m/n_m)/f_m}{-\ln(y_s/n_s)/f_s}$$

and the log mutant frequency (*LMF*) is simply

$$LMF = \ln(MF)$$

The heterogeneity factor should be used to modify the estimate of variance. Again using the notation from Box 4.1 and defining H_s and H_m to be the heterogeneity factors for survival and mutation, the variance (V) of the *LMF* becomes

$$V = \frac{H_s(1-y_s/n_s)}{y_s(\ln(y_s/n_s))^2} + \frac{H_m(1-y_m/n_m)}{y_m(\ln(y_m/n_m))^2}$$

The statistical weight (W) of *MF* is required for the estimation of trend. It is

$$W = 1/(V(MF)^2)$$

Worked examples of these calculations are shown in Box 4.4.

Box 4.4. *Worked example of calculation of mutant frequency and log mutant frequency and of the variance of log mutant frequency and the weight of mutant frequency*

Calculation of mutant frequency (MF) for dose 0 (data from Boxes 4.1 and 4.2)

$$MF_0 = \frac{-\ln(y_m/n_m)/f_m}{-\ln(y_s/n_s)/f_s}$$

$$= \frac{-\ln(1441/1533)/4000}{-\ln(68/384)/2.0}$$

$$= 1.788 \times 10^{-5}$$

Calculation of log mutant frequency (LMF) for dose 0 (data from Boxes 4.1 and 4.2)

$$LMF_0 = \ln(MF_0)$$

$$= -10.932$$

Calculation of the variance (V) of LMF, using the new estimates of the heterogeneity factors from Box 4.3.

$$V_0 = \frac{H_{s\ new}(1 - y_s/n_s)}{y_s(\ln(y_s/n_s))^2} + \frac{H_{m\ new}(1 - y_m/n_m)}{y_m(\ln(y_m/n_m))^2}$$

$$= \frac{1.758(1 - 68/384)}{68(\ln(68/384))^2} + \frac{2.914(1 - 1441/1533)}{1441(\ln(1441/1533))^2} = 0.0388$$

Calculation of the weight (W) of MF

$$W = \frac{1}{V(MF)^2}$$

$$W_0 = \frac{1}{0.0388(1.788 \times 10^{-5})^2}$$

$$= 8.062 \times 10^{10}$$

Result of calculations for all doses

Dose	MF	W	LMF	V
0	1.788×10^{-5}	8.062×10^{10}	-10.932	0.0388
1	1.795×10^{-5}	4.200×10^{10}	-10.928	0.0739
2	3.511×10^{-5}	1.319×10^{10}	-10.257	0.0615
4		excluded		
8	5.462×10^{-5}	7.601×10^{9}	-9.815	0.0441
16	8.243×10^{-5}	4.420×10^{9}	-9.404	0.0333

4.2.3.2 *Comparison of control with each treatment dose*

Furth *et al.* (1981) use the variance of the LMF for the control to compute an upper 95% (one-sided) confidence limit so that any treatment yielding a value above this is considered significant. This procedure ignores the uncertainty in estimating the LMF for the treatment and consequently will erroneously declare too many treatments to be significant. As Thomas

Box 4.5. *Worked example of comparison of each treatment with the control*

For each treatment calculate

$$D_i = LMF_i - LMF_c$$
$$var(D_i) = V_i + V_c$$

and the statistic $\frac{D_i^2}{var(D_i)}$ is calculated

For dose 16, using values from Box 4.4.

$$D_i = (-9.404) - (-10.932)$$
$$= 1.528$$
$$var(D_i) = 0.0333 + 0.0388$$
$$= 0.0721$$
$$\frac{D_i^2}{var(D_i)} = \frac{1.528^2}{0.0721}$$
$$= 32.38$$

Calculation for all doses

Dose	D_i	$var(D_i)$	$D_i^2/var(D_i)$
1	0.004	0.1127	0.00014
2	0.675	0.1003	4.54
4	excluded		
8	1.117	0.0829	15.05
16	1.528	0.0721	32.38

Use Table 4.4. to test for significance.
Comparing 4 doses with the control:

$D_i^2/var(D_i)$ for doses 8 and 16 is greater than 4.67

(personal communication) implies, the correct method is to calculate the lower 95% confidence limit of the difference, treatment *LMF* (LMF_i) minus control *LMF* (LMF_C), and if it is greater than 0 the treatment effect is considered significant. This corrected approach is not as convenient to apply as that of Furth *et al.* since the variances of the treatment *LMF*s differ.

We recommend that the equivalent significance test be used in place of the confidence limits. The procedure is set out with a worked example in Box 4.5. The difference

$$D_i = LMF_i - LMF_c$$

is found together with its variance,

Table 4.4. *Comparison of several treatments with the same control (from one-sided Dunnett's test)*

	Number of treatments, excluding the control (i.e. number of comparisons)							
	1	2	3	4	5	6	7	8
$P<5\%$	2.71	3.69	4.24	4.67	4.97	5.24	5.48	5.66

Source: adapted from Dunnett (1955).

$$var(D_i) = V_i + V_c,$$

where V_i and V_c are the variances of the treatment and control *LMF*s.

If D_i is greater than 0, the test statistic

$$\frac{D_i^2}{var(D_i)}$$

is computed. This is compared to the critical values in Table 4.4, which is based on Dunnett's test for multiple comparisons with the same control (Dunnett, 1955). If the relevant critical value is exceeded there is a significant difference between the control and the treatment. The critical values are approximations for the application described here because they are derived on the assumption of equal variances for the control and treatment. When the control variance is smaller than the treatment variance, the actual probability is slightly greater than that given in Table 4.4, and when the control variance is larger, the actual probability is slightly less.

4.2.3.3 *Test for linear trend of mutant frequency with dose*

The simplest realistic model of linear trend is of the form:

$$MF = SMF + b\cdot\text{dose}$$

where *MF* is the observed mutant frequency, *SMF* the spontaneous mutant

Table 4.5. *Critical values of the χ^2 distribution with one degree of freedom (one-sided) (equivalent to the square of the normal distribution)*

Probability	5%	1%	0.1%
χ^2	2.71	5.41	9.55

Source: from Pearson & Hartley (1976).

Box 4.6. *Worked example of a test for linear trend of mutant frequency with dose*

Using the values for Mutant Frequency (MF) and its weight (W) in Box 4.4, calculate the following:

The sum of the weights

$$\begin{aligned} SW &= \sum[W] \\ &= 8.062\times10^{10}+4.200\times10^{10}+1.319\times10^{10} \\ &\quad +7.601\times10^{9}+4.420\times10^{9} \\ &= 1.478\times10^{11} \end{aligned}$$

The sum of weight × mutant frequency

$$\begin{aligned} SWM &= \sum[W\times MF] \\ &= (8.062\times10^{10})(1.788\times10^{-5})+(4.200\times10^{10})(1.795\times10^{-5}) \\ &\quad +(1.319\times10^{10})(3.511\times10^{-5})+(7.601\times10^{9})(5.462\times10^{-5}) \\ &\quad +(4.420\times10^{9})(8.243\times10^{-5}) \\ &= 3.438\times10^{6} \end{aligned}$$

The sum of weight × dose

$$\begin{aligned} SWD &= \sum[W\times D] \\ &= (8.062\times10^{10})0+(4.200\times10^{10})1+(1.319\times10^{10})2 \\ &\quad +(7.601\times10^{9})8+(4.420\times10^{9})16 \\ &= 1.999\times10^{11} \end{aligned}$$

The sum of weight × dose²

$$\begin{aligned} SWDD &= \sum[W\times D^2] \\ &= (8.062\times10^{10})(0^2)+(4.200\times10^{10})(1^2)+(1.319\times10^{10})(2^2) \\ &\quad +(7.601\times10^{9})(8^2)+(4.420\times10^{9})(16^2) \\ &= 1.713\times10^{12} \end{aligned}$$

The sum of weight × dose × mutant frequency

$$\begin{aligned} SWDM &= \sum[W\times D\times M] \\ &= (8.062\times10^{10})0(1.788\times10^{-5})+(4.200\times10^{10})1(1.795\times10^{-5}) \\ &\quad +(1.319\times10^{10})2(3.511\times10^{-5})+(7.601\times10^{9})8(5.462\times10^{-5}) \\ &\quad +(4.420\times10^{9})16(8.243\times10^{-5}) \\ &= 1.083\times10^{7} \end{aligned}$$

Next calculate

$$\begin{aligned} S1 &= SWDM - SWD\times SWM/SW \\ &= 1.083\times10^{7}-(1.999\times10^{11})(3.438\times10^{6})/(1.478\times10^{11}) \\ &= 6.18\times10^{6} \\ S2 &= SWDD-(SWD)^2/SW \\ &= 1.713\times10^{12}-(1.999\times10^{11})^2/(1.478\times10^{11}) \\ &= 1.443\times10^{12} \end{aligned}$$

The slope b and its variance, $var(b)$, are

$$\begin{aligned} b &= S1/S2 \\ &= (6.18\times10^{6})/(1.443\times10^{12}) \\ &= 4.28\times10^{-6} \end{aligned}$$

and

Box 4.6. (*cont.*)

$$var(b) = \frac{1}{S2}$$

Finally calculate the test statistic, $b^2/var(b)$

$$= \frac{(S1/S2)^2}{(1/S2)} = S1^2/S2$$
$$= (6.18 \times 10^6)^2/(1.443 \times 10^{12})$$
$$= 26.47$$

Since this is greater than 9.55 (Table 4.5), $P < 0.001$

frequency and b is the slope which can be estimated, using weighted regression. The weights are obtained from the variances of *LMF* by

$$W = \frac{1}{V(MF)^2}$$

as shown in Section 4.2.3.1.

Box 4.6 illustrates how to calculate the slope b and its variance $var(b)$ to form the test statistic $b^2/var(b)$ which should be compared with tabulated critical values of χ^2 with 1 degree of freedom in Table 4.5 (i.e. with the square of the standard normal distribution). A one-sided test should be used since only a positive value of the slope b is of interest. If the 5% critical value is exceeded there is evidence of an increase in mutant frequency with increase in dose.

More sophisticated models might be advanced, and might have relevance to particular systems, but it is our belief that they would add little in the general case.

4.2.3.4 *Checking consistency across experiments*

More than one experiment should always be carried out to compare a substance with a control, possibly utilising different doses of the test substance. It is then desirable to pool information from the several experiments in order to obtain a more powerful test for mutagenicity. The experiments should of course relate to the same conditions, for example with regard to metabolic activation.

Prior to combining information from experiments it is advisable to check that the experiments are reasonably consistent. To do this, the test statistic is again distributed as χ^2 and it is computed for, say, the ith treatment as

$$\sum[D_j^2/var(D_j)] - (\sum[D_j/var(D_j)])^2/\sum[1/var(D_j)]$$

Table 4.6. *Test for consistency (χ^2 test, two-sided)*

	Number of trays or experiments						
	2	3	4	5	6	7	8
Between trays (0.1%)	10.8	13.8	16.3	18.5	20.5	22.5	24.3
Between experiments (1%)	6.6	9.2	11.3	13.3	15.1	16.8	18.5

Source: from Pearson & Hartley (1976).

Box 4.7. *Comparison of each treatment with control. Worked example of the test for consistency between experiments and of the method of combining results from separate experiments*

Consistency of comparison of control with each treatment dose:
Values have been obtained (see Box 4.5) for

$$D_j = LMF_i - LMF_c$$

where D is the difference between treatment and control for the ith treatment of the jth experiment,
and for its variance

$$var(D_j) = V_i + V_c$$

	Expt 1		Expt 2	
Dose	D	$var(D)$	D	$var(D)$
1	0.004	0.1127	0.126	0.1103
2	0.675	0.1003	0.408	0.1066
4	—	—	0.720	0.0984
8	1.117	0.0829	1.344	0.0780
16	1.528	0.0721	1.641	0.0739

Next calculate for each dose, (e.g. for dose 1)

$S = \sum[1/V_j]$ $(= 1/0.1127 + 1/0.1103 = 17.939)$
$SD = \sum[D_j/V_j]$ $(= 0.004/0.1127 + 0.126/0.1103 = 1.178)$
$SD2 = \sum[D_j^2/V_j]$ $(= (0.004)^2/0.1127 + (0.126)^2/0.1103 = 0.1441)$
$SD3 = SD2 - (SD)^2/S$ $(= 0.1441 - (1.178)^2/17.939 = 0.067)$

Comparing $SD3$ with critical values in Table 4.6, the results at this dose are consistent and can thus be combined.
Combining results from experiments:
The combined estimate of the difference between control and treatment is (example for dose 1)

$$CD = SD/S \ (= 1.178/17.939 = 0.0657)$$

with variance

$$var(CD) = 1/S$$

Box 4.7. (*cont.*)

and the test statistic

$$(CD)^2/var(CD) = (SD/S)^2/(1/S) = (SD)^2/S$$
$$= (1.178)^2/17.939 = 0.0774)$$

Results of these calculations for all doses are

Dose	S	SD	$SD2$	$SD3$	CD	$(SD)^2/S$
1	17.939	1.178	0.1441	0.067	0.0657	0.0774
2	19.351	10.557	6.104	0.345	0.546	5.77
4	—	—	—	—	—	—
8	24.883	30.705	38.209	0.320	1.234	37.89
16	27.401	43.398	68.822	0.088	1.584	68.74

Comparing $SD3$ with critical values in Table 4.6, the results at each dose are consistent and can be combined. Testing the combined difference (CD) for significance, the critical value for 4 doses from Table 4.4 is 4.67 so that the effects for doses 2, 8 and 16 are significant.

where

D_j is $LMF_i - LMF_c$
$var(D_j) = V_i + V_c$ for the jth experiment

Compare the test statistic with the χ^2 table with $(M-1)$ degrees of freedom where there are M experiments. If the 1% level, tabulated in Table 4.6, is exceeded, then reporting a combined result would be inappropriate. The 1% level is suggested to allow for multiple testing which otherwise would produce too many apparently inconsistent experiments by chance alone.

A worked example of the above statistic is set out in Box 4.7, together with the procedure for combining results from separate experiments.

The corresponding statistic for the trend test is similar, having D_j replaced by b and $var(D_j)$ by $var(b)$ and the worked example is set out in Box 4.8.

4.2.3.5 *Combining information from experiments*

Considering first the comparison of the control and the ith treatment, the combined estimate of the difference between the control and a treatment is:

$$CD = \frac{\sum[D_j/var(D_j)]}{\sum[1/var(D_j)]}$$

where j extends over the M experiments and $var(D_j) = V_c + V_i$.
If $CD > 0$ its variance should be calculated as

Box 4.8. *Test for trend. Worked example of the test for consistency between experiments and of the method for combining data from several experiments*

Experiment 1 $b = 4.28 \times 10^{-6}$

$$S2 = 1/var(b) = 1.443 \times 10^{12}$$

(from Box 4.6)

Experiment 2 $b = 3.46 \times 10^{-6}$

$$S2 = 1/var(b) = 9.249 \times 10^{11}$$

Calculate

$$\begin{aligned}
S &= \sum[S2] \\
&= 1.443 \times 10^{12} + 9.429 \times 10^{11} \\
&= 2.386 \times 10^{12} \\
SD &= \sum[bS2] \\
&= (4.28 \times 10^{-6})(1.443 \times 10^{12}) + (3.46 \times 10^{-6})(9.249 \times 10^{11}) \\
&= 9.376 \times 10^{6} \\
SD2 &= \sum[b^2 S2] \\
&= (4.28 \times 10^{-6})^2(1.443 \times 10^{12}) + (3.46 \times 10^{-6})^2(9.249 \times 10^{11}) \\
&= 37.506 \\
SD3 &= SD2 - (SD)^2/S \\
&= 37.506 - (9.376 \times 10^{6})^2/(2.386 \times 10^{12}) \\
&= 0.662
\end{aligned}$$

which is less than the critical value of 6.6 for two experiments, obtained from Table 4.6, so that the two experiments can be combined.

Combining experiments:

The combined estimate of the slope (CS) is

$$\begin{aligned}
CS &= SD/S \\
&= (9.376 \times 10^{6})/(2.386 \times 10^{12}) \\
&= 3.930 \times 10^{-6}
\end{aligned}$$

with variance

$$var(CS) = 1/S$$

and the test statistic is

$$(CS)^2/var(CS) = (SD)^2/S = (9.376 \times 10^{6})^2/(2.386 \times 10^{12}) = 36.84$$

Since this is greater than 9.55 (Table 4.5), $P < 0.001$.

$$var(CD) = \frac{1}{\sum[1/var(D_j)]}$$

and the test statistic $CD^2/var(CD)$ compared with the critical values in Table 4.4.

The trend test can be pooled over experiments in like manner, replacing D_j by b and $var(D_j)$ by $var(b)$ and entering Table 4.5 if the combined slope is greater than 0.

Boxes 4.7 and 4.8 show worked examples of these tests.

4.2.4 Miscellaneous matters relating to the heterogeneity factor

4.2.4.1 *Initial estimate of the heterogeneity factor*

About 50 to 100 pairs of cultures should be used in the initial estimate of the factor. Clearly, these will not have featured in a single experiment but in many. Laboratories already performing the mammalian cell fluctuation test are likely to have pertinent data from past experiments. Combining data in this way is justified on the assumption that no changes have occurred in the test procedure. It would be inappropriate to include data collected during the period when the test was being introduced, either to the laboratory as a whole or to individual staff. Until a laboratory has gathered sufficient information to permit estimates of its own, we suggest ours (1.8 for survival and 3.0 for mutation, see Section 4.2.2.1) be used as a starting point. This would be more realistic than ignoring the factors altogether.

4.2.4.2 *Quality control of test performance*

In some ways the size of the heterogeneity factor might be regarded as an inverse measure of the quality or integrity of the assay procedure since, as explained in Section 4.2.2, it is directly related to the extra variability, beyond the theoretical value from the binomial distribution, introduced during the course of the test. Whilst it certainly should not be employed as a definitive judge of the performance of a laboratory it should convey some information of this nature.

Our current values of 1.8 and 3.0 for survival and mutation trays respectively show what levels can be achieved. They have been derived from data provided by experienced laboratories. It should therefore be cause for concern if factors much larger than these are encountered. The methods and procedures should be examined to ascertain possible explanations. Similarly, if it is found that the calculated factors are deteriorating (increasing) with time then explanations should be sought.

4.2.4.3 *Checking consistency of trays*

So far it has been assumed that, whenever more than one tray has been used for a culture or condition or treatment/control combination, then the results on the individual trays within a replicate treatment were sufficiently alike to justify summing over the trays. It is likely that this assumption will be true on most occasions, but nevertheless it should always be formally checked.

Suppose there are M trays and that on the ith tray there is a total of n_i wells of which y_i show no growth.

Box 4.9. *Worked example of a test for consistency between trays. Using data from individual trays pooled to give replicate 1 dose 1 in Box 4.1*

Tray M	negative wells y	total wells n	fraction negative p
1	94	96	0.9792
2	89	96	0.9271
3	91	96	0.9479
4	88	96	0.9167
	$Y=362$	$N=384$	$P=0.9427$

Calculate

$$\sum[y_i^2/n_i]$$
$$=94^2/96+89^2/96+91^2/96+88^2/96$$
$$=341.479$$

$$X^2=\frac{N(N\sum[y_i^2/n_i]-Y^2)}{Y(N-Y)}$$
$$=\frac{384(384\times 341.479-362^2)}{362(384-362)}$$
$$=4.047$$

Compare with the critical value (16.3) for $M=4$ trays in Table 4.6. Since the present value of 4.047 is lower, the trays are consistent and can be summed.

Let

$$N=\sum[n_i]$$

and

$$Y=\sum[y_i]$$

Then compute the test statistic

$$X^2=\frac{N(N\sum[y_i^2/n_i]-Y^2)}{Y(N-Y)}$$

X^2 should be compared with the chi-squared table with $(M-1)$ degrees of freedom. Sets of trays with X^2 greater than the critical value for significance at the 0.1% level should be omitted from the experiment. These values are tabulated in Table 4.6. Box 4.9 shows a worked example of this procedure, which is the variance test for homogeneity of the binomial distribution as described, for example, in Snedecor & Cochran (1967).

If more than two sets of trays have to be rejected for this reason it is advisable to repeat the whole experiment, after determining the cause of the excessive variability between trays.

The formula for X^2 simplifies when all the n_i are identical (i.e. there are the same number of wells on each tray) to become:

$$X^2 = \frac{N(M\sum[y_i^2] - Y^2)}{Y(N-Y)}$$

which further simplifies when there are only two trays to:

$$X^2 = \frac{N(y_1 - y_2)^2}{Y(N-Y)}$$

4.2.4.4 *Determination of the heterogeneity factor from more than two cultures*

When there are more than two replicate cultures the determination of the heterogeneity factor becomes more complex. The calculations, though, are almost the same as those in Section 4.2.4.3.

Suppose for the control or treatment in a given set of conditions there are M cultures and that for the ith one there is a total of n_i wells of which y_i show no growth. (Note that the same nomenclature is being used as in Section 4.2.4.3 but that the n_i and y_i are now the sums over the replicate trays for the ith culture.)

Let $N = \sum[n_i]$ and $Y = \sum[y_i]$.

Then compute

$$X^2 = \frac{N(N\sum[y_i^2/n_i] - Y^2)}{Y(N-Y)}$$

The formula for X^2 is identical to that in Section 4.2.4.3 with the changed definition of n_i and y_i, and it simplifies similarly when all the n_i are equal.

The heterogeneity factor is estimated as

$$h = X^2/(M-1)$$

with $(M-1)$ degrees of freedom.

This estimate should be compared with the current factor to guard against the occurrence of technical errors. The ratio of this estimate to the current estimate should rarely exceed a critical value, dependent on the degrees of freedom $(M-1)$. The values in Table 4.3 should be used. These stem from the F-test at the one-sided 0.1% level.

We recommend that groups of cultures which yield excessively high estimates should be excluded from analysis. The weighted mean of all the *h*s obtained under the conditions of interest gives the heterogeneity factor for the experiment. This reduces to

$$h_{exp} = \frac{\sum[X_j^2]}{\sum[M_j - 1]}$$

where the *j*th set has M_j cultures yielding X_j^2.

The estimate from the experiment should be checked against the current factor. The largest allowable ratio of the two depends on the total degrees of freedom for the estimate, namely $\sum[M_j - 1]$. In the common case where the control and all treatments have the same number of cultures, M, the degrees of freedom are $C(M-1)$, there being C sets of cultures. See Table 4.3 for the critical ratio, using the one-sided 1% level. If this ratio is exceeded the experiment should be discarded.

4.3 BACTERIAL TESTS

In bacterial fluctuation tests, the bacteria are incubated in the continuous presence of the test agent. This makes it impractical to determine survival, and the test is for an absolute increase in the proportion of wells containing mutants. Since there is no direct test for survival, it is necessary to test a wider range of doses, in order to obtain results with some doses close to the limit of toxicity. Bacterial tests are likely to involve the use of five or more test strains, with or without the *S*9 metabolic activation system, giving ten or more tests, only one or two of which may be expected to give positive results, perhaps over a narrow dose range, with any given mutagenic test compound. Hence it is essential that any interpretation makes allowance for multiple comparisons.

4.3.1 Sampling distribution

Since in a bacterial test there is continuous incubation with the test compound, there is no subculturing and the additional error associated with subculturing will not be introduced. Our investigations have provided no evidence to invalidate the assumption of a binomial sampling distribution for the variation between replicate trays. Hence this assumption has been made in our recommended methods of analysis of data from the bacterial fluctuation test.

Nevertheless, the assumption of a binomial distribution should be verified for each experiment, and replicate trays should be used for the control and preferably one or more treatment doses. Replicates should of course be prepared as separate treatments, not merely by splitting the same mixture into separate trays. The mechanics of testing for heterogeneity have been described in Section 4.2.2.1 for the mammalian cell test and a heterogeneity factor can be calculated in exactly the same way. For well-conducted bacterial tests, however, the factor should be little greater than one. Nevertheless, since binomial variance is the theoretical minimum variance possible in a fluctuation test, ignoring the factor will mean that the actual variance is likely to be somewhat greater than that assumed in the statistical test and this will lead to a slight excess of false positives. The advice with regard to excessive variability is to omit a set of trays if so indicated but to abandon the whole experiment if the control set or more than one treatment set shows significantly greater than binomial variation.

We suggest that a heterogeneity factor be calculated and updated using the procedure described in Sections 4.2.2 and 4.2.4. An appropriate starting point for updating the bacterial heterogeneity factor would be 1.0. Monitoring the factor should be of assistance in quality control and in evaluating borderline data.

4.3.2 Recommended analyses

4.3.2.1 *Comparison of control and treatment concentrations*

As with the mammalian cell test in Section 4.2.3.2 we recommend that each treatment be compared individually with the control in the bacterial test. Dunnett's multiple comparison test is utilised to allow for repeated significance testing.

Data for the control and a single dose of the treatment can be laid out in the form of a 2 × 2 table, after the counts have been summed over individual trays when there is replication. In the notation of Box 4.10.

$$\chi^2 = \frac{(N-1)(n_i y_c - n_c y_i)^2}{n_c n_i Y(N-Y)}$$

Values of χ^2 can be compared to those in Table 4.4 for an approximate version of Dunnett's test. The factor $(N-1)$ in the above expression replaces the usual one of N for theoretical reasons (removing bias), as advocated by Gilbert (1980), although such modification has negligible

Table 4.7. *Example of bacterial fluctuation test data*

i (dose)	0	1	2	3	4	5	
Negative wells y_i	157	82	72	71	72	74	$Y=\sum y_i=528$
Total wells n_i	192	96	96	96	96	96	$N=\sum n_i=672$
$v_i=y_i/n_i$	0.817	0.854	0.750	0.740	0.750	0.771	

Box 4.10. *Form of bacterial fluctuation test data and a worked example of a test of an individual dose against the control*

	Wells negative	Wells positive	Total wells
Control	y_c	$n_c - y_c$	n_c
Treated	y_i	$n_i - y_i$	n_i
Total	Y	$N - Y$	N

Using the treatment giving the greatest number of positive wells from Table 4.7

	Wells negative	Wells positive	Total wells
Control	157	35	192
Treated	71	25	96
Total	228	60	288

$$\chi^2 = \frac{(N-1)(n_i y_c - n_c y_i)^2}{n_c n_i Y(N-Y)}$$

$$= \frac{(288-1)(96 \times 157 - 192 \times 71)^2}{192 \times 96 \times 228 \times 60}$$

$$= 2.36$$

which is <4.97, the 5% critical value for the comparisons of five treatments with the same control (from Table 4.4) and hence not significant.

effect since N is large. If the heterogeneity factor calculated for the bacterial test is greater than 1, the value of χ^2 should be divided by the factor before entering Table 4.4. If the heterogeneity factor is less than 1, it should be ignored.

Both Gilbert (1980) and Collings *et al.* (1981) pointed out that, since only an increase in the number of positive wells in the treatment over the control is of interest, a one-sided test should be employed. Therefore χ^2 should be computed only if $(n_i y_c - n_c y_i) > 0$.

When there are equal numbers of wells used for the control and the treatment doses, namely $N/2$, the above expression simplifies to

$$\chi^2 = \frac{(N-1)(y_c - y_i)^2}{Y(N-Y)}$$

which should be computed only if $(y_c - y_i) > 0$.

Each dose should be compared with the control in turn. A single significant difference should not be regarded as firm evidence of mutagenicity. A consistent result (see Section 4.3.2.3), though, over the same dose range in two or preferably three independent experiments should be evidence of mutagenicity (Hubbard *et al.*, 1984). When results are equivocal our advice echoes that of Hubbard, that experiments should be repeated 'until a consistent picture emerges'. Sections 4.3.2.3 and 4.3.2.4 outline statistical procedures for checking consistency across experiments and for subsequently combining information from experiments.

4.3.2.2 *Test for trend*

Simpson & Margolin (1986) recently investigated the characteristics of various tests for dose-response relationships with a toxic effect at high doses. They warn that without careful attention to the decision rules for excluding data beyond the apparent peak mutagenic response, the reported significance levels can be seriously understated. In the bacterial fluctuation test, a dose-related increase in wells containing mutants (and corresponding decrease in negative wells) would be evidence of a mutagenic effect as in the mammalian cell test. However a clear dose response is often unobtainable since toxicity can mask such an effect and since the results in the bacterial fluctuation test are not adjusted for survival.

The performance of the test recommended here (Collings *et al.*, 1981) is unlikely to be affected badly when a number of treatments have used a level of test agent too low to show any effect. On the other hand, the performance is likely to suffer when the highest doses are acutely toxic since the method would require these to be combined with lower doses, which might well help to conceal a genuine mutagenic effect at a lower dose. We suggest that the test be modified to exclude such doses as follows: The dose giving the highest frequency of mutants (i.e. the lowest frequency of negative wells) should be determined. Each higher dose should be compared with this maximum frequency to identify the first one giving significantly fewer mutants (i.e. more negative wells). These comparisons should be made using the statistic and Dunnett's test set out in Section 4.3.2.1 with the negative control replaced by the dose giving the most mutants and testing for a significant reduction in the number of mutants. When Table 4.4 is entered, the appropriate number of comparisons will be the number of doses higher than that giving the maximum frequency of mutants. If a dose does yield a significantly lower result then it and all higher doses should be excluded from the test of trend.

Collings *et al.* (1981) examined three tests for trend. They strongly preferred the so-called isotonic test to others considered, because of its robustness in the presence of toxicity. In such circumstances the number of negative wells reaches a minimum with increasing dose, and then increases, severely reducing the power of the standard linear regression approach described in Section 4.2.3.3 to detect the initial trend.

Isotonic (or order-preserving) regression involves fitting values to the observed proportions of negative wells for each dose such that they either decrease or remain the same with each increase in dose. The control can be (and should be) included in the regression. The test for isotonic trend is performed on those dose levels not excluded as toxic as described above.

A method of obtaining the maximum likelihood estimates of the fitted values is the 'Pool the Adjacent Violators' algorithm described in Barlow *et al.* (1972). For the ith dose let the current fitted proportion of wells negative be v_i, then the v_i are adjusted until, for all i, $v_i >= v_j$ where $i<j$. At the start of the procedure v_i is set equal to y_i/n_i, where y_i negative wells were observed out of n_i, summed over trays if relevant. Consecutive doses are grouped into blocks for which v_i is made equal to $\sum y_i/\sum n_i$ where the sum extends over those doses in the block. The smallest number of blocks which satisfies the restrictions yields the isotonic solution.

The procedure, which is iterative, is set out in Box 4.11, using the data in Table 4.7.

At the beginning of a cycle, the lowest i is found for which $v_i < v_j, j=i+1$, and then the block containing v_i is pooled (amalgamated) with the following block and the vs are revised. In the example, $v_0 < v_1$ so the first two blocks (equal to the doses at this point) are pooled, giving their summed $y_{01}=239$, $n_{01}=288$, and revised $v_0=v_1=v_{01}=0.83$. The new enlarged block is compared with the next one, and these two are pooled if necessary, which is not in this example since $v_{01}=0.83>v_2=0.75$. If not, the higher doses are compared similarly. We have $v_2(0.75)>v_3(0.74)$ as desired, but $v_3(0.74)<v_4(0.75)$. Blocks 3 and 4 are pooled to give $y_{34}=143$, $n_{34}=192$, and revised $v_3=v_4=v_{34}=0.745$. As $v_{34}(0.745)<v_5(0.77)$ it is necessary to pool the three doses 3, 4 and 5, to form a single block with $y_{345}=217$, $n_{345}=288$, and revised $v_3=v_4=v_5=v_{345}=0.753$.

When the top dose has been considered the process is repeated; this continues until no blocks are changed in a cycle, showing the required solution has been obtained, given by the v_is. In the second cycle we see $v_2(0.75)<v_{345}(0.753)$, so adding dose 2 to the end block leads to revised $v_2=v_3=v_4=v_5=v_{2345}=289/384=0.7526$. These calculations are laid out in Box 4.11.

When the sample sizes are equal, that is when $n_i=n$ for all i, the pooling can be done in terms of the y_is directly.

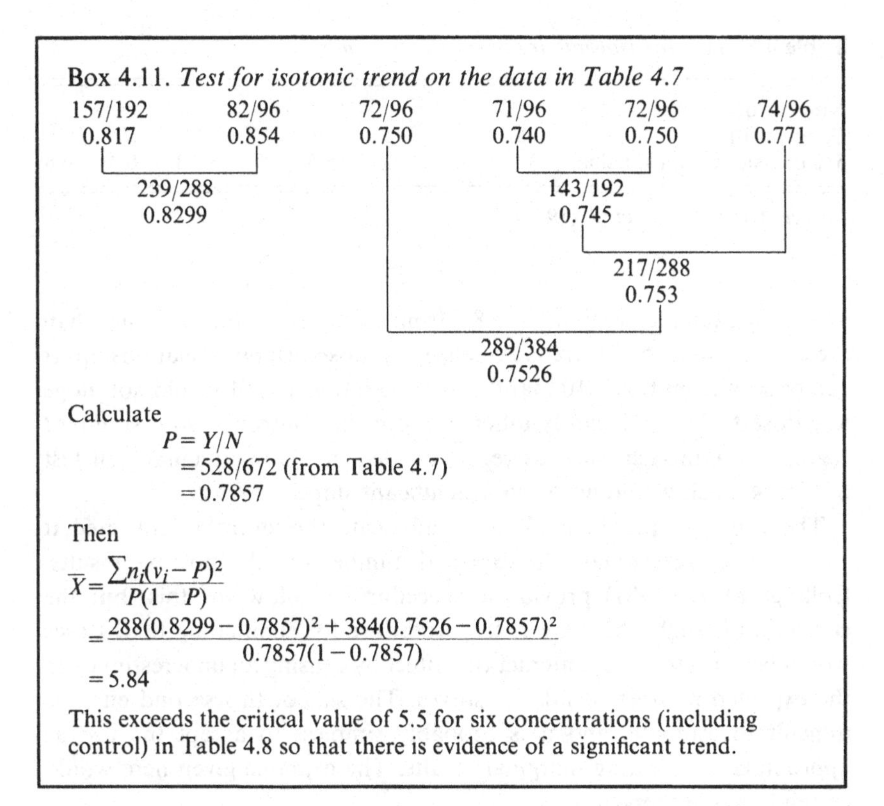

Box 4.11. *Test for isotonic trend on the data in Table 4.7*

157/192	82/96	72/96	71/96	72/96	74/96
0.817	0.854	0.750	0.740	0.750	0.771

239/288
0.8299

143/192
0.745

217/288
0.753

289/384
0.7526

Calculate

$$P = Y/N$$
$$= 528/672 \text{ (from Table 4.7)}$$
$$= 0.7857$$

Then

$$\bar{X} = \frac{\sum n_i(v_i - P)^2}{P(1-P)}$$
$$= \frac{288(0.8299 - 0.7857)^2 + 384(0.7526 - 0.7857)^2}{0.7857(1-0.7857)}$$
$$= 5.84$$

This exceeds the critical value of 5.5 for six concentrations (including control) in Table 4.8 so that there is evidence of a significant trend.

The statistic for testing the fit to the observed data is similar in form to that in the usual χ^2 test for goodness of fit.

$$\bar{X} = \frac{\sum n_i(v_i - P)^2}{P(1-P)}$$

where $P = \sum[y_i]/\sum[n_i] = Y/N$

For equal sample sizes,

$$\bar{X} = \frac{kN\sum[(y_i - Y/k)^2]}{Y(N-Y)}$$

where there are k doses, including the control, using n wells.

The distribution of $\bar{X}$ is not straightforward but tables exist for the case of equal samples sizes. Table 4.8, extracted from Table A3 of Barlow *et al.* (1972), gives the one-sided 5% critical values for up to ten doses (including the control) in an experiment. If $\bar{X}$ exceeds the appropriate entry then there is evidence of a dose response relationship.

Table 4.8. *Test for isotonic trend. Critical values*

Number of Concentrations	3	4	5	6	7	8	9	10
5% one-sided critical value	3.8	4.5	5.0	5.5	5.8	6.1	6.3	6.6

Source: from Barlow *et al.* (1972).

For the example, we find $\overline{X} = 5.84$, from six doses. As this is greater than the critical value, 5.5, there is evidence of a dose-response relationship. It can be seen from Box 4.10 that the test of Section 4.3.2.1 would not judge any dose to be significantly different from the control. Also a standard regression approach, such as recommended in the mammalian cell test Section 4.2.3.3, would not yield a significant slope.

The $\overline{X}$ test is approximate. If, as we advocate, the control is replicated, it will slightly overestimate the expected number of false positive results. Collings *et al.* (1981) provide a procedure to allow for this, but the correction is slight. Moreover, the procedure for eliminating toxic doses would be expected to counteract this effect by causing an underestimate in the expected number of false positives. The size of this second effect is difficult to estimate, and it is probably simplest to accept the test as approximate, repeating marginal results. The example given here would certainly require repeating.

4.3.2.3 *Checking consistency across experiments*

As with the mammalian cell test, consistency across experiments can be checked statistically and then, if the check is satisfactory, the results can be combined. The method, due to Mantel & Haenzel (Fleiss, 1973), is as follows for the comparison of a treatment with the control:

Compute the sums over the M experiments:

$$SX2 = \sum [x^2]$$

$$S1 = \sum \left[\frac{(n_c y_i - n_i y_c)}{N} \right]$$

$$S2 = \sum \left[\frac{n_c n_i Y(N-Y)}{N^2(N-1)} \right]$$

Then compare $SX2 - (S1)^2/S2$ with the 1% chi-squared critical values with $(M-1)$ degrees of freedom in Table 4.6. If the tabulated value is exceeded there is evidence that the experiments are not consistent with each other, so that combining them would not be justified. This test should be carried out for each treatment.

We are unable to offer a consistency check for isotonic regression. However, if any of the individual treatment and control comparisons that have not been excluded for toxicity fail the above test, then it would be inadvisable to attempt to combine the experiments for the trend test.

4.3.2.4 *Combining information from experiments*

For the comparison of the control with a treatment, the sums $S1$ and $S2$ defined in the previous section are used. The test statistic $(S1)^2/S2$ should be compared with the critical values in Table 4.4, derived from Dunnett's test, but only if $S1 > 0$ since a one-sided test is appropriate. The test should of course be carried out for each treatment.

For an overall test of trend we suggest that the data simply be summed over the experiments for the control and each dose and the procedures described in Section 4.3.2.2. be applied.

4.4. CONCLUSIONS

(a) We *recommend* including *genuinely independent replicates* of each treatment in a mammalian fluctuation test. Without these, it is not possible to perform a valid test of significance. With bacterial tests, there should be independent replicates of the control and preferably also of some of the treatments.

(b) We make a number of other recommendations concerning experimental design, appropriate numbers of cells, etc.

(c) We *recommend* the calculation of a *heterogeneity factor* for mammalian fluctuation tests, allowing historical data to be used to improve the estimate of variance between independent replicates of the same treatment. For bacterial tests we *suggest* that a similar factor be calculated, as a form of quality control and to assist in decision making.

(d) For both mammalian and bacterial data, we describe a test for comparison of the control with each treatment dose. For mammalian data, weighted mean log mutant frequencies are calculated and compared using an estimate of variance modified by the heterogeneity factor. For bacterial data, a simpler 2×2 test can be performed on the actual data.

(e) For mammalian data we describe a test for linear trend using weighted regression. Because viability is not determined for bacterial data, a test for isotonic trend is described, which reduces problems associated with toxicity.

(f) We *recommend* that all experiments be repeated. The repeat

Table 4.9. *Recommended sequence of steps in the analysis of mammalian fluctuation tests*

	See Section
A For each experiment	
1. For each culture check consistency of trays	4.2.4.3
2. For control and each treatment	
(a) compute heterogeneity factor	4.2.2.1 and 4.2.4.4
(b) compare heterogeneity factor with current value	4.2.2.1 and 4.2.4.4
3. For survival and mutation separately	
(a) find average heterogeneity factor	4.2.2.1 and 4.2.4.4
(b) compare average heterogeneity factor with current value	4.2.2.2. and 4.2.4.4
(c) combine average heterogeneity factor with current value	4.2.2.3
4. For control and each treatment compute *MF*, *LMF*, variance of *LMF* and weight of *MF*	4.2.3.1
5. For each treatment	
(a) compute difference in *LMF* from control *LMF*, and its variance	4.2.3.2
(b) test difference for significance	4.2.3.2
6. Over all treatments test for linear trend of *MF* with dose	4.2.3.3
B After two or more experiments	
1. For each relevant treatment comparison of *LMF* with control *LMF*	
(a) check consistency between experiments	4.2.3.4
(b) combine comparisons and test for significance	4.2.3.5
2. Over all treatments	
(a) check consistency of linear trend between experiments	4.2.3.4
(b) combine estimates of trend and test for significance	4.2.3.5

experiments should be tested for consistency and, except with the test for isotonic trend, where they are consistent the combined data should be tested for significance.

(g) Tables 4.9 and 4.10 outline the sequences of steps in the analysis of mammalian and bacterial data respectively.

(h) Finally, we would suggest that the conclusion drawn from one of these statistical tests constitutes a prediction of the likely result of a similar experiment in the future. If it is not possible to reach a clear decision from the available data, the matter should be resolved by doing one or more additional experiments.

4.5 REFERENCES

Barlow, R.E., Bartholomew, D.J., Bremner, J.M. & Brunk, H.D. (1972). *Statistical Inference under Order Restrictions.* Wiley, New York, pp. 13–16.

Table 4.10. *Recommended sequence of steps in the analysis of bacterial fluctuation test data*

	See Section
A For each experiment	
1. For control and each treatment	
(a) compute heterogeneity factor	4.2.2.1 and 4.2.4.4
(b) compare heterogeneity factor with current value	4.2.2.1 and 4.2.4.4.
2. For control and all treatments together	
(a) find average heterogeneity factor	4.2.2.1 and 4.2.4.4
(b) compare average heterogeneity factor with current value	4.2.2.2 and 4.2.4.4
(c) combine average heterogeneity factor with current value	4.2.2.3
3. Compare each treatment with control	4.3.2.1
4. Over all treatments	
(a) eliminate toxic treatments	4.3.2.2 and 4.3.2.1
(b) test significance of isotonic regression	4.3.2.2
B After two or more experiments	
1. For each relevant treatment comparison with control	
(a) check consistency between experiments	4.3.2.3
(b) combine comparisons and test for significance	4.3.2.4
2. Over all treatments	
(a) combine the data over experiments	4.3.2.4
(b) eliminate toxic treatments	4.3.2.2
(c) check consistency between experiments via procedure of 1.(a)	4.3.2.3
(d) test significance of isotonic regression	4.3.2.4

Cole, J. & Arlett, C.F. (1984). The detection of gene mutations in cultured mammalian cells. In: *Mutagenicity Testing – A Practical Approach*, Ed. S. Venitt & J.M. Parry, IRL Press, Oxford, pp. 233–73.

Cole, J., Arlett, C.F. & Green, M.H.L. (1976). The fluctuation test as a more sensitive system for determining induced mutation in L5178Y mouse lymphoma cells. *Mutation Research*, **41**, 377–86.

Cole, J., Arlett, C.F., Green, M.H.L., Lowe, J. & Muriel, W. (1983a). A comparison of the agar cloning and microtitration techniques for assaying cell survival and mutation frequency in L5178Y mouse lymphoma cells. *Mutation Research.*, **111**, 371–86.

Cole, J., Fox, M., Garner, C., McGregor, D. & Thacker, J. (1983*b*). Gene mutation assays in cultured mammalian cells. In: *UKEMS Sub-committee on Guidelines for Mutagenicity Testing*. Report. Part 1: Basic test battery. Ed. B.J. Dean, United Kingdom Environmental Mutagen Society, Swansea, pp. 65–102.

Collings, B.J., Margolin, B.H. & Oehlert, G.W. (1981). Analyses for binomial data, with application to the fluctuation test for mutagenicity. *Biometrics*, **37**, 775–94.

Dunnett, C.W. (1955). A multiple comparison procedure for comparing several treatments with a control. *Journal of American Statistical Society*, 1096–120.

Finney, D.J. (1978). *Statistical Method in Biological Assay*. 3rd edn, Griffin, London, p. 373.

Fisher, R.A. (1954). *Statistical Methods for Research Workers*. Oliver & Boyd, Edinburgh, pp. 61–2.

Fleiss, J.L. (1973). *Statistical Methods for Rates and Proportions*. Wiley, New York, p. 117.

Furth, E.E., Thilly, W.G., Penman, B.W., Liber, H.L. & Rand, W.M. (1981). Quantitative assay for mutation in diploid human lymphoblasts using microtiter plates. *Analytical Biochemistry*, **110**, 1–8.

Gilbert, R.I. (1980). The analysis of fluctuation tests. *Mutation Research*, **74**, 283–9.

Green, M.H.L., Muriel, W.J. & Bridges, B.A. (1976). Use of a simplified fluctuation test to detect low levels of mutagens. *Mutation Research*, **38**, 33–42.

Hubbard, S.A., Green, M.H.L., Gatehouse, D. & Bridges, J.W. (1984). The fluctuation test in bacteria. In: *Handbook of Mutagenicity Test Procedures*, Ed. B.J. Kilbey, M. Legator, W. Nichols & C. Ramel, Elsevier, Amsterdam.

Lea, D.E. & Coulson, C.A. (1949). The distribution of numbers of mutants in bacterial populations. *Journal of Genetics*, **49**, 264–85.

Leong, P.-M., Thilly, W.G. & Morgenthaler, S. (1985). Variance estimation in single-cell mutation assays: comparison to experimental observations in human lymphoblasts at 4 gene loci. *Mutation Research*, **150**, 403–10.

Luria, S.E. & Delbruck, M. (1943). Mutations of bacteria from virus sensitivity to virus resistance. *Genetics*, **28**, 491–511.

Pearson, E.S. & Hartley, H.O. (1976). *Biometrika Tables for Statisticians*, Vol. 1, 3rd edn, Griffin, London.

Simpson, D.G. & Margolin, B.H. (1986). Recursive nonparametric testing for dose-response relationships subject to downturns at high doses. *Biometrika*, **73**, 589–96.

Snedecor, G.W. & Cochran, W.G. (1967). *Statistical Methods*, 6th edn. Iowa University Press, Iowa, p. 240.

Thilly, W.G., DeLuca, J.G., Furth, E.E., Hoppe IV, H., Kaden, D., Krolewski, J.J., Liber, H.L., Skopek, T.R., Slapikoff, S.A., Tizard, R.J. & Penman, B.W (1980). Gene-locus mutation assays in diploid human lymphoblast lines. In: *Chemical Mutagens*, vol. 6, ed. F.J. de Serres & A. Hollaender, Plenum, New York pp. 331–64.

Venitt, S. (1982). UKEMS collaborative genotoxicity trial. Bacterial mutation tests for 4-chloromethylbiphenyl, 4-hydroxymethylbiphenyl and benzyl chloride. Analysis of data from 17 laboratories. *Mutation Research*, **100**, 91–109.

5

Analysis of data from *in vitro* cytogenetic assays

* C. RICHARDSON (Group Leader)
* D. A. WILLIAMS	J. A. ALLEN
G. AMPHLETT	D. O. CHANTER
B. PHILLIPS

* Co-authors

5.1 INTRODUCTION

Methodology for the performance of *in vitro* cytogenetic assays was described by Scott *et al.* (1983), and full details of appropriate protocols, cell types, preparation of test material and factors influencing the performance of the assays are discussed in that report. Detailed statistical treatments were not discussed and little or no additional guidance is provided in regulatory guidelines, where the most commonly used phrase is that 'appropriate statistical methods should be employed'. In this chapter we recommend an appropriate statistical method for the analysis of data from *in vitro* cytogenetic assays.

Section 5.2. considers the various factors of experimental design and chromosomal analysis which defined the form and structure of the data available for statistical analysis. A recommended method of statistical analysis is developed in Sections 5.3–5.5, and Section 5.6 discusses more technical statistical issues and mentions alternative statistical analyses. Section 5.7 gives a number of recommendations and a number of steps that may be taken when evaluating cytogenetic data.

5.2 BIOLOGICAL FACTORS AFFECTING THE FORM AND STRUCTURE OF DATA

5.2.1 The assay

Two cell types may be used for the generation of *in vitro* cytogenetic data. These include primary cell cultures, e.g. human lymphocytes and established (i.e. immortalised) cell lines such as Chinese hamster fibroblasts. Whichever cell type is used the following experimental design features are desirable for the provision of acceptable data.

(a) Blank and solvent controls should be used.
(b) Positive control cultures should be used to establish that the system is sensitive to both direct and indirect acting mutagens.
(c) A minimum of three dose levels of test chemical should be used so as to define adequately any dose-response relationship observed.
(d) Duplicate cultures should be set up for each dose level, with preferably an increased replication of the solvent controls.
(e) The assay should be conducted in the presence and absence of auxiliary metabolic activation.

Other factors which may be less critical in statistical terms are discussed by Scott *et al.* (1983).

5.2.2 Selection of cells for chromosomal analysis

Scott *et al.* (1983) recommended scoring a minimum of 200 cells for each treatment implying the scoring of 100 cells per culture if duplicate cultures are set up for each dose level. However, it may be preferable to increase the replication of the solvent control to four cultures, as this improves the sensitivity of the assay. Scoring 100 cells per culture may not always be possible and the implications are discussed in Section 5.3. Cells with chromosome numbers close to the normal range (in the case of established cell lines) or of at least $2n-2$ are acceptable, and techniques must be used that ensure the same cell is not scored more than once.

5.2.3 Assessment of chromosomal damage

The end-point of the majority of *in vitro* cytogenetic assays is the observation of chromosomal damage. These observations are usually made on metaphase cells but can be performed by the counting of anaphase bridges and fragments, or micronuclei in interphase cells. (These two methods have not yet assumed the popularity of metaphase analysis in *in vitro* studies and are not considered further in this chapter.) The types of damage observed in metaphase cells can be divided into two categories, i.e. structural chromosomal aberrations and numerical alterations.

The most widely used assays involve the scoring of structural aberrations such as breaks, fragments, exchange figures etc and achromatic gaps. The inclusion or exclusion of this last category of chromosomal damage has been at the heart of a great deal of debate over the past few years. At present, most regulatory guidelines recommend that gaps should be recorded but not included in the calculation of the percentage of abnormal cells. It is common practice in many laboratories to conduct statistical analysis on aberrations or aberrant cells including and excluding gaps. The methods discussed in this chapter can be used whether or not gaps are taken into account when classifying a cell as abnormal. The scoring of numerical aberrations, although employed in a number of assays, is more problematical, as events such as hypodiploidy may be induced as artefacts of preparative technique (e.g. too much hypotonic treatment). However, it is realised that some laboratories do employ assays that have the identification of aneuploidy as an end-point, and the methods of statistical analysis described later are appropriate for such an end-point.

5.2.4 Selection of dose level of test chemical

The choice of dose levels should be made after consideration of at least one of the following factors:

(a) suppression of mitotic index
(b) growth inhibition
(c) cell death
(d) colony forming ability
(e) the limit of solubility of the compound.

In general, the most commonly used method is to identify doses which produce a reduction in mitotic index. The OECD (1983) guideline states 'The highest test substance concentration with and without an exogenous metabolic activation system should suppress mitotic activity by approximately 50%. Relatively insoluble substances should be tested up to the limit of solubility. For freely soluble non-toxic substances the upper test substance concentration should be determined on a case-by-case basis'. Once the top dose level has been defined, it is normal, when using the assay for screening purposes, to use two further dose levels, for example at $\frac{1}{2}$ and $\frac{1}{10}$ of the top dose. However, if limited clastogenic activity is observed (e.g. at a single dose), it may be necessary to use a narrower dose range and/or more dose levels, in order to define a dose-response relationship.

5.2.5 Sampling time

When evidence exists to suggest that the test compound is inducing mitotic delay, additional harvest times should be used. In some instances it may be useful to use 5-bromodeoxyuridine to determine the optimal harvest time. When multiple harvesting is undertaken each sampling time should be treated as a single experiment for statistical evaluation, although it may be appropriate to pool data from control cultures at each time point for these statistical comparisons.

5.2.6 Recording and presentation of chromosomal aberration data

In general, the cell types used in *in vitro* cytogenetic studies have a very low prevalence of each aberration type, and therefore a meaningful statistical method will normally require the grouping of aberration types. A suitable scheme for recording and presenting chromosomal aberration data is described in detail by Scott *et al.* (1983).

Summary data are usually presented as percentage of cells with aberrations or number of aberrations/cell. However, a small increase in an aberration type with exceedingly low background frequency, e.g. exchange figures (Richardson *et al.* 1984) may constitute a biologically significant event; in these cases it may be advisable to undertake serial sampling studies to maximise aberration yields.

It is the opinion of this group that the cell, rather than the individual chromatid or chromosome, is regarded as the base unit for assessment of damage. The reasons for this point of view are: first, the usual consequence of chromosomal damage is that the cell will die, irrespective of the number of aberrations in that cell (Brewen, 1982). Thus, once one aberration has occurred in a cell, additional aberrations have little or no further biological relevance. Secondly, the frequency of multiple aberration is usually extremely low (except for very potent clastogens), so very little extra information is provided by distinguishing between single and multiple aberrations. If, in practice, it was important to know whether clustering of aberrations had occurred, the frequency distribution of the numbers of aberrations per cell would have to be constructed for each treatment group, and analysed by some appropriate technique.

We appreciate that some investigators will choose to analyse data presented as aberrations per cell. The methods described in this chapter are also considered to be appropriate methods for the analysis of data displayed in this manner.

5.2.7 Pre-requisites of data analysis

It is essential that the following points are checked before a statistical analysis is performed.

(a) Data for individual solvent control cultures fall within an acceptable range and are compatible with historical data. If this is not the case a repeat experiment should be performed.

(b) The positive control substances induce sufficient aberrations to confirm effectiveness of the test procedures.

Table 5.1. *Data summaries from three assays, Sets 1, 2 and 3. Exactly 100 cells were sampled from each culture*

Treatment	Number of cultures (m_i)	Number of aberrant cells in each culture (r_{ij})	Total number of cells sampled (n_i)	Total aberrant (r_i)	Proportion aberrant (p_i)
Set 1					
Negative control	4	0, 2, 5, 1	400	8	0.020
Dose 1 200 μg/ml	2	2, 1	200	3	0.015
Dose 2 1000 μg/ml	2	6, 3	200	9	0.045
Dose 3 2000 μg/ml	2	7, 9	200	16	0.080
Positive control	2	18, 54	200	72	0.360
Set 2					
Negative control	4	2, 1, 0, 0	400	3	0.0075
Dose 1 20 μg/ml	2	4, 1	200	5	0.025
Dose 2 100 μg/ml	2	6, 8	200	14	0.070
Dose 3 200 μg/ml	2	3, 1	200	4	0.020
Positive control	2	10, 14	200	24	0.120
Set 3					
Negative control	4	1, 0, 3, 1	400	5	0.0125
Dose 1 62.5 μg/ml	2	0, 2	200	2	0.010
Dose 2 125 μg/ml	2	1, 1	200	2	0.010
Dose 3 250 μg/ml	2	0, 4	200	4	0.020
Dose 4 500 μg/ml	2	4, 3	200	7	0.035
Positive control	2	27, 28	200	55	0.275

5.3 PRELIMINARY STATISTICAL CONSIDERATIONS

The data to be analysed are, for each culture, the number of cells sampled and the number of cells exhibiting at least one aberration. Three sets of data are displayed in Table 5.1. Set 1 is from an assay using Chinese hamster ovary cells while Sets 2 and 3 are from assays using human lymphocytes. In these assays, all from the same laboratory, exactly 100 cells were sampled from each culture. It is not always possible in practice to achieve exactly equal sample sizes, and any recommended method of analysis must be sufficiently flexible to permit variation in the number of cells sampled from each culture.

Note that in all these data sets it is apparent without formal statistical analysis that the positive control has increased the incidence of aberrations relative to the negative control, so each assay is evidently capable of detecting a clastogenic compound. Once this has been established the positive control data play no further role in the statistical analysis.

The experimental design implies that the sample of cells from the same culture should be regarded as an experimental unit, and, in assessing the evidence for differences between treatments, the variation between treatments should be compared with the variation within treatments between replicate cultures. In a carefully executed assay we might hope to achieve homogeneity between replicate cultures so that, within each treatment, the cells sampled from different cultures are no more variable than cells sampled from the same culture. If such homogeneity is achieved then the number of aberrant cells will vary between replicate cultures according to a binomial distribution. The assumption of homogeneity can therefore be tested using the binomial dispersion test.

Some notation is needed to describe this test. We number the treatments $0, 1, \ldots, t$ where 0 is the negative control and $1, \ldots, t$ are the test treatments. In the ith treatment $(i=0, 1, \ldots, t)$ the number of replicate cultures is m_i, from the jth $(j=1, 2, \ldots, m_i)$ of which n_{ij} cells are sampled of which r_{ij} are found to contain aberrations. Summing over the replicate cultures in the ith treatment the total number of cells sampled is n_i of which r_i are aberrant. The proportion of aberrant cells in the ith treatment is $p_i = r_i/n_i$. Thus, for data Set 1, $t=3$, $m_0=4$, $m_1=m_2=m_3=2$, $n_{ij}=100$ for all i and j, $r_{01}=0$, $r_{02}=2$ etc., $n_0=400$, $r_0=8$ and $p_0=8/400=0.02$. The binomial dispersion test compares the test statistic

$$X^2 = \sum_{i=0}^{t} \sum_{j=1}^{m_i} \frac{(r_{ij}-n_{ij}p_i)^2}{n_{ij}p_i(1-p_i)}$$

with the upper percentage points of the χ^2 distribution with $\sum_{i=0}^{t} (m_i-1)$ degrees of freedom. (Note that X^2 denotes the chi-squared test statistic while χ^2 denotes the chi-squared distribution.) Applying this test to data Set 1 we obtain

$$X^2 = \frac{(-2)^2+0^2+3^2+(-1)^2}{1.96} + \frac{(0.5)^2+(-0.5)^2}{1.4775} + \frac{1.5^2+(-1.5)^2}{4.2975}$$
$$+ \frac{(-1)^2+1^2}{7.36} = 7.143+0.338+1.047+0.272 = 8.80$$

The upper 5% point of the χ^2 distribution with 6 *d.f.* is 12.59, the upper 20% point is 8.56, so there is no evidence here of heterogeneity between cultures. Note however that there is substantial heterogeneity between the two positive control replicates. An alternative heterogeneity test is described in Section 5.6.1.

We have examined the variation between replicate cultures in a large number of data sets from three laboratories. In negative controls, and in test treatments which appear to have little or no effect compared with negative controls, we found that the numbers of aberrant cells varied between replicates to an extent that is compatible with binomial variation. Only in positive controls and in test treatments which exhibit a marked increase in the proportion of aberrant cells was there evidence of heterogeneity between replicate cultures. The findings are compatible with those of Margolin *et al.* (1986) who concluded from their study of experimental data that homogeneous binomial variation is an achievable goal.

The statistical tests of treatment differences proposed in Section 5.4 will therefore assume that, in negative controls and test treatments which do not increase the incidence of aberrations, the variation between replicate cultures is binomial. This implies that the amount of information on the response to any treatment depends on the total number of cells sampled, irrespective of the number of replicate cultures. There is no statistical advantage in sampling, say, 20 cells from ten replicates, rather than 100 cells from two replicates. Of course, it is possible that data from some laboratories will exhibit heterogeneity between replicates. In practice, the variation between replicates should always be inspected. If there is evidence of heterogeneity then the method of analysis described below will not be valid and statistical advice will need to be sought. If homogeneity cannot be achieved, then increased replication of cultures within treatments will be required in order to estimate the extra-binomial variation and so obtain a valid measure of the precision of treatment differences. In fact the situation becomes comparable to *in vivo* cyto, enetic assays, with the *in vitro* variation between replicate cultures being analogous to the *in vivo* variation between replicate animals.

5.4 COMPARISON OF TEST TREATMENTS WITH NEGATIVE CONTROL

Assuming binomial variation between replicate cultures the response of cells to treatment i is characterised by a single parameter θ_i, which is the probability that a randomly chosen cell is aberrant. The total number r_i of aberrant cells in any treatment i has a binomial distribution with expectation $n_i\theta_i$. If the test agent has no effect then $\theta_0 = \theta_1 = \ldots = \theta_t$. If the test agent increases the incidence of aberrations at dose i then $\theta_0 < \theta_i$. We can therefore consider each test treatment proportion of aberrant cells, p_i, in turn and compare it with the negative control proportion, p_0, using a test of the hypothesis $\theta_0 = \theta_i$ against the one-sided alternative $\theta_0 < \theta_i$.

Approximate tests using the normal approximation to the binomial distribution could be somewhat unreliable because the frequencies r_i are small. We therefore prefer to use Fisher's exact test. For dose i, putting $N=n_0+n_i$ and $R=r_0+r_i$, the formula for calculating the significance probability is

$$P=h(0)+\ldots+h(r_0-1)+\tfrac{1}{2}h(r_0),$$

where $h(x)$ denotes the hypergeometric probability $h(x)=\binom{n_0}{x}\binom{n_i}{R-x}\Big/\binom{N}{R}$. Here $\binom{N}{R}$ denotes the binomial coefficient $\dfrac{N(N-1)\ldots(N-R+1)}{R(R-1)\ldots 1}$ and $\binom{n_0}{x}$ and $\binom{n_i}{R-x}$ are similarly defined. Reasons for choosing this definition of the significance probability, rather than the more usual $P_+=h(0)+\ldots+h(r_0)$, are given in Section 5.6.3.

When writing a computer program to calculate P the simplest approach is to create an array $L(x)=\ln(x!)$, $x=0, 1, \ldots, N$. The hypergeometric probabilities $h(x)$ are then given by $h(x)=\exp[L(n_0)+L(n_i)+L(R)+L(N-R)-L(x)-L(n_0-x)-L(R-x)-L(n_i-R+x)-L(N)]$.

For data Set 1 we obtain the following significance probabilities.

$$\text{Dose 1: } h(x)=\binom{400}{x}\binom{200}{11-x}\Big/\binom{600}{11}, P=h(0)+\ldots+h(7)+\tfrac{1}{2}h(8)=0.648,$$

$$\text{Dose 2: } h(x)=\binom{400}{x}\binom{200}{17-x}\Big/\binom{600}{17}, P=h(0)+\ldots+h(7)+\tfrac{1}{2}h(8)=0.049,$$

$$\text{Dose 3: } h(x)=\binom{400}{x}\binom{200}{24-x}\Big/\binom{600}{24}, P=h(0)+\ldots+h(7)+\tfrac{1}{2}h(8)=0.0004,$$

Given the very strong evidence that $\theta_3>\theta_0$, and the weaker evidence that $\theta_2>\theta_0$, we conclude that the test agent is clastogenic.

In this example the data are compatible with a monotonically increasing dose-response curve (i.e. a curve whose slope is never negative). Scott *et al.* (1983, §4.4) warn that the dose–response curve may not be monotonic. The shape of the dose–response curve depends on the sampling time. It is possible for the incidence of aberrations to increase to a maximum at some intermediate dose and then to decrease over higher doses. Scott *et al.* (1983, §4.4) suggest that a test agent should be judged clastogenic if there are significant increases at two consecutive doses. Certainly care is needed when interpreting results in which there is a significant increase at only one dose, for when multiple comparisons are made with a control, there is a greater than 5% chance that at least one of the doses will produce an increase significant at the 5% level. This multiple comparisons problem can be avoided by using a single test, combining the responses at different doses, as suggested in Section 5.6.2. Instructions on how to calculate P using a standard computer program for the Fisher exact test are also given in that section.

5.5 ANALYSIS OF DATA SETS

5.5.1 Set 1

These data are analysed in Sections 5.3 and 5.4 above.

5.5.2 Set 2

The binomial dispersion test statistic is 6.87 with 6 *d.f.*, so there is no evidence of heterogeneity between replicate cultures. Comparing each dose with the negative control using Fisher's exact test we obtain significance probabilities of $P=0.053$ for dose 1, $P=0.00002$ for dose 2, $P=0.108$ for dose 3. Although there is evidence of an increase only at dose 2 (100 μg/ml) the significance probability is extremely small and justifies the conclusion that the test agent is probably clastogenic, but that further investigation is advisable.

In a repeat assay using a much smaller range of doses centred around 100 μg/ml the following proportions of aberrant cells were recorded.

Negative control	0/100, 1/100, 4/100, 5/100
Dose 1 60μg/ml	11/100, 21/100
Dose 2 80μg/ml	27/114, 14/86
Dose 3 100μg/ml	13/56, 33/113
Dose 4 120μg/ml	27/106, 2/31
Dose 5 140μg/ml	5/27, 15/57

It is now evident, without formal statistical analysis, that the test agent is clastogenic. Note that in the test treatments here there is evidence of heterogeneity between cultures.

This example demonstrates that a clastogenic compound may increase the incidence of aberrations significantly at only one dose level if the doses are too widely spaced. If there is evidence of an effect at only one dose level it is prudent to carry out a repeat assay using a smaller range of doses.

5.5.3 Set 3

The binomial dispersion test statistic is 10.10 with 7 *d.f.* so the assumption of binomial variation between replicates is acceptable. Comparison with the negative control using Fisher's exact test gives significance

Table 5.2. *The homogeneity test statistic* $-2\sum ln\ P_i$ *calculated for Data set 1*

Comparison i	Treatment	Proportions compared	Two-sided P value	$-2\ln P_i$
1	Control	0/100 v 2/100	0.2487	2.783
2	Control	5/100 v 1/100	0.1203	4.235
3	Control	2/200 v 6/200	0.1758	3.477
4	Dose 1	2/100 v 1/100	0.6231	0.946
5	Dose 2	6/100 v 3/100	0.3337	2.195
6	Dose 3	7/100 v 9/100	0.6153	0.971
				Total = 14.607

probabilities $P=0.58$ at doses 1 and 2, $P=0.25$ at dose 3 and $P=0.041$ at dose 4. There is therefore some evidence of an increase in aberrations at the highest dose, but these data alone are not sufficient to justify the conclusions that the test agent is clastogenic. In such situations further information can be obtained by scoring additional cells from the same assay. If a clear conclusion does not emerge from an analysis of the enhanced data a repeat assay should be conducted.

5.6 DISCUSSION

We have described one way in which *in vitro* chromosome aberration data can be analysed. The analysis uses two well-established statistical tests: the binomial dispersion test and Fisher's exact test. This analysis is simple given access to a computer or programmable calculator for the calculation of Fisher's exact test. If, however, a statistician is responsible for analysing the data he or she may well choose to use alternative, less well known, test procedures. An alternative test of homogeneity between replicate cultures is described in Section 5.6.1, while an alternative approach to testing treatment differences is outlined in Section 5.6.2. Finally Section 5.6.3 discusses the definition of significance probability adopted in Section 5.3.

5.6.1 An alternative test of homogeneity between replicate cultures

In Section 5.3 the binomial dispersion test statistic was referred to a χ^2 distribution. The χ^2 distribution approximation may not be reliable when the numbers of aberrant cells observed per replicate are small. An alternative approach is to divide the cultures into separate duplicate pairs and use two-sided Fisher's exact tests to test the homogeneity of each duplicate pair. The two-side significance probability here is the minimum of $2P$ and $2(1-P)$, where P is the one-sided significance probability defined in Section 5.4. The set of two-sided significance probabilities, $P_1, P_2, \ldots, P_k$ say, from the comparison of k duplicate pairs, are then combined into a single overall test by comparing $-2\sum \ln P_i$ with a χ^2 distribution with $2k$ degrees of freedom. The calculation of $-2\sum \ln P_i$ for data Set 1 is detailed in Table 5.2. Note that the four negative control cultures provide three independent pairwise comparisons, the third, 2/200 *v* 6/200, being a comparison of the first two cultures with the last two cultures. The statistic $-2\sum \ln P_i = 14.61$ is not significantly large compared with the χ^2 distribution with 12 *d.f.*

5.6.2 Alternative tests of treatment differences

In Section 5.4 the evidence for treatment differences was assessed by comparing each test treatment with the negative control. It may be preferable to perform a single test using a statistic chosen to be sensitive to the kind of overall dose response pattern that the cytogeneticist aims primarily to detect. This approach could be more powerful, and it avoids the difficulty of combining the results of multiple comparisons. Margolin *et al.* (1986) suggest that the Cochran-Armitage trend test should be used. This test would be insensitive to dose response patterns like those in Set 2, in which the incidence of aberrations reaches a maximum at some intermediate dose. Test statistics can be devised which will detect an increase in the incidence of aberrations without assuming that the maximum increase is achieved at the highest dose. The test statistic does not need to be a simple function of the data with a known approximating distribution under the null hypothesis, for the significance probability can be evaluated by using a computer to enumerate, or sample from, the exact permutation distribution. For a fuller development of these ideas see Williams (1988). The assistance of an experienced statistician will be needed to implement such test procedures.

5.6.3 The definition of significance probability for Fisher's exact test

In Section 5.4 the significance probability P of the one-sided Fisher exact test was defined as $P = h(0) + \ldots + h(r_0 - 1) + \frac{1}{2}h(r_0)$. Thus $P = \frac{1}{2}(P_- + P_+)$ where $P_- = h(0) + \ldots + h(r_0 - 1)$, $P_+ = h(0) + \ldots + h(r_0)$. The definition of significance probability adopted by most statisticians is P_+, but both P_- and P_+ provide relevant information. The value of P_+ implies that a test using a significance level $\alpha \geqslant P_+$ would reject the hypothesis being tested, while the value of P_- implies that a test with a significance level $\alpha \leqslant P_-$ would accept the hypothesis. Because the reponse variable is discrete we cannot actually carry out a test with a significance level intermediate between P_- and P_+ (unless we use a randomised test which, though theoretically valid, is in practice unacceptable). It is helpful to know the value of both P_- and P_+. For example the values $P_- = 0.02$, $P_+ = 0.07$ would cast more doubt on the hypothesis being tested than the values $P_- = 0.06$, $P_+ = 0.07$. If only a single value is to be quoted as a measure of the weight of evidence against the null hypothesis, then we favour the mid-*P* value $P = \frac{1}{2}(P_- + P_+)$. One advantage that *P* has over P_+ is that the distribution of *P* under the null hypothesis is closer to that of the uniform distribution and it has an exact expectation of 0.5. This is particularly important when several *P* values are combined as in Section 5.6.2. Other arguments favouring the mid-*P* value are given by Lancaster (1961), Anscombe (1981, p. 289), Franck (1986) and Haber (1986).

The *P*-value as defined here can easily be derived from computer programs which give the more usual P_+ value. For example, comparing Dose 2 with the negative control in data Set 1 the usual comparison of 9/200 with 8/400 gives $P_+ = 0.0726$. P_- is obtained by assuming one more aberrant cell in the treated group and one fewer in the control group; thus comparing 10/200 with 7/400 gives $P_- = 0.0255$. The average of these two is the required mid-*P* value, 0.049.

5.7 RECOMMENDATIONS ON THE ASSESSMENT OF DATA

This chapter has discussed and finally recommended a statistical method for the evaluation of *in vitro* cytogenetic data. However, the investigator should be mindful of other considerations that should be made when evaluating data, e.g. the application of good scientific judgment. An attempt to demonstrate a possible scheme for data interpretation following the statistical analysis is shown below.

(a) No statistically significant increase in the number of aberrant cells (at any dose level) above concurrent control frequencies — conclusion NEGATIVE.

(b) An increase in the number of aberrant cells, at least at one dose level, which is so substantially greater than the laboratory negative control range that formal statistical analysis is deemed unnecessary – conclusion POSITIVE.

(c) A statistically significant increase in the number of aberrant cells, at least at one dose level, which is not substantially greater than the laboratory negative control range — EVALUATE FURTHER.

(d) An increased incidence of unique or rare types of aberration — EVALUATE FURTHER.

Further evaluation may comprise one or more of the following:

(a) Comparison of the levels of chromosomal aberrations with historical control frequencies.

(b) Extended scoring (i.e. more cells) of affected cultures, and assessment of higher dose levels if available.

(c) Repeating the appropriate part of the experiment with the possible use of a higher and/or narrower dose range, and/or serial sampling.

Once (a), (b) and (c) above have been done, interpretation is based on the previously stated criteria.

5.8 SUMMARY

(a) A typical *in vitro* cytogenetic assay compares a solvent control, three dose levels of the test chemical and a positive control, using at least two replicate cultures per treatment. From each culture a sample of at least 100 cells is scored and each structural chromosome aberration is categorised and recorded.

(b) The statistical analysis proposed here is appropriate after first reducing the information on each cell to a binary classification, each cell being classified as normal or aberrant. The experimenter has freedom to ignore certain categories of chromosome damage (e.g. achromatic gaps) when making this classification.

(c) In well-conducted assays there should be no extra-binomial variation between the proportions of aberrant cells in replicate cultures. This should be confirmed by using a binomial dispersion test.

(d) Assuming that the variation between replicate cultures is binomial, the replicates can be combined to give a single proportion of aberrant cells for each treatment. Each treatment proportion is then compared with the negative control proportion using a Fisher exact test. Scientific judgment is needed when interpreting the results of these tests.

(e) Alternative statistical test procedures are available which take a more coherent view of the whole dose-response pattern, but these require greater statistical expertise.

5.9 REFERENCES

Anscombe, F.J. (1981). *Computing in Statistical Science Through APL*, Springer Verlag, New York.

Brewen, G. (1982). Chromosome aberrations in mammals as genetic parameters in determining mutagenic potential and assessory genetic risk. *Progress in Mutation Research* Vol. 3, Eds K.C. Bora *et al.* Elsevier Biomedical Press.

Franck, W.E. (1986). P-values for discrete test statistics. *Biometrical Journal* **28**, 403–6.

Haber, M. (1986). A modified exact test for 2 × 2 contingency tables. *Biometrical Journal* **28**, 455–63.

Lancaster, H.O. (1961). Significance tests in discrete distributions. *Journal of the American Statistical Association* **56**, 223–34.

Margolin, B.H., Resnick, M.A., Rimpo, J.Y., Archer, P., Galloway, S.M., Bloom, A.D. & Zeiger, E. (1986). Statistical analysis for *in vitro* cytogenetic assays using Chinese hamster ovary cells. *Environmental Mutagenesis* **8**, 183–204.

OECD (1983). Guidelines for Testing of Chemicals. *Genetic Toxicology: In Vitro Mammalian Cytogenetic Tests. Guideline 473.* Organisation for Economic Cooperation and Development, Paris.

Richardson, C.R., C.A. Howard., T. Sheldon, J. Wildgoose & M.G. Thomas (1984). The human lymphocyte *in vitro* cytogenetic assay: positive and negative control observations on ≈30,000 cells. *Mutation Research*, **141**, 59–64.

Scott, D., Danford, N., Dean, B., Kirkland, D. & Richardson, C. (1983). *In vitro* chromosome aberration assays, in *UKEMS Sub-Committee on Guidelines for Mutagenicity Testing*, Report, Part 1. Basic Test Battery. Ed. B.J. Dean, United Kingdom Environmental Mutagen Society, Swansea, pp. 41–64.

Williams, D.A., (1988). Test for differences between several small proportions. *Applied Statistics* **37**, 421–434.

6

Statistical methods for sister chromatid exchange experiments

D COOKE J. ALLEN M.G. CLARE
C.J. DORÉ L. HENDERSON (Group Leader)

6.1 INTRODUCTION

Sister chromatid exchanges (SCE) are reciprocal exchanges between sister chromatids. They are generally visualised by exposing cells (*in vitro* or *in vivo*) to 5-bromodeoxyuridine for two cell cycles, allowing subsequent differential staining of sister chromatids. Exchanges are detected as switches in stained regions between sister chromatids. The molecular mechanism of SCE is not known but they occur after exposure to many genotoxic agents and are believed to indicate DNA damage. Consequently several SCE assays have been developed to assess the genotoxic potential of chemicals. The assays involve the exposure of either cultured cells or experimental animals to chemicals under test, followed by exposure to 5-bromodeoxyuridine for a sufficient period to allow two rounds of DNA synthesis. The number of SCEs per metaphase and the number of chromosomes in each metaphase are scored.

Guidelines for the conduct of these tests have been published by regulatory authorities and learned societies (Organisation for Economic Co-operation and Development (OECD, 1986), US Environmental Protection Agency (EPA, 1983), and UK Environmental Mutagen Society (UKEMS), (Perry *et al.*, 1984)). Whilst much attention has been paid to the methodology of these tests, appropriate statistical methods have received less attention. The relevant OECD and EPA guidelines simply recommend 'data should be evaluated by appropriate statistical methods'. Methods of statistical analysis for some individual assay systems have been analysed in detail, e.g. the Chinese hamster ovary (CHO) SCE assay (Margolin *et al.*, 1986). In this chapter we present a generalised approach to the statistical analysis of SCE. Data from three basic assay types have been used: (a) *in vitro* assay using the CHO cell line, (b) *in vitro* assay using human lymphocytes and (c) exposure of animals *in vivo*.

Table 6.1. *SCE counts from eight cultures of Chinese hamster ovary cells, consisting of two cultures at each of four doses (0, 1, 3.3 and 10 μg/ml) of a compound*

Dose (μg/ml)	Culture	SCE counts per cell										Total
0	1	8	4	6	6	12	10	5	3	8	8	70
	2	8	5	9	7	10	6	3	5	6	9	68
1	3	11	8	12	2	7	11	4	8	7	13	83
	4	9	8	13	6	7	3	7	10	6	6	75
3.3	5	6	8	6	6	1	6	11	7	4	5	60
	6	8	7	9	7	6	9	10	9	7	6	78
10	7	10	7	4	11	9	11	8	5	7	7	79
	8	10	10	6	6	11	10	8	8	13	5	87

Note: For the purpose of illustration, ten cells have been selected at random from the 50 cells scored for SCE for each culture.

6.2 EXPERIMENTAL DESIGN

6.2.1 Introduction

The experimental unit in SCE experiments is either a culture (*in vitro* experiments) or an animal (*in vivo* experiments), since this is the unit to which treatments are applied at random. Many cells are scored for SCE for each experimental unit. The significance of treatment effects is assessed by comparison with the variation between cultures (animals), often called 'the error variation', or briefly, 'the error'.

SCE experiments usually have a simple hierarchical structure. Each of the t treatments is applied to r cultures, and each culture contains n cells (see Fig. 6.1). Tables 6.1 and 6.2 give data from experiments with Chinese hamster ovary cells (CHO) and mouse haemopoeitic cells, respectively. These data sets will be used in examples.

The variation between all the observations in an experiment may be regarded as being derived from three sources: (1) between treatments, (2) between cultures within treatments, (3) between cells within cultures. This

Fig. 6.1. Hierarchical structure of experiments.

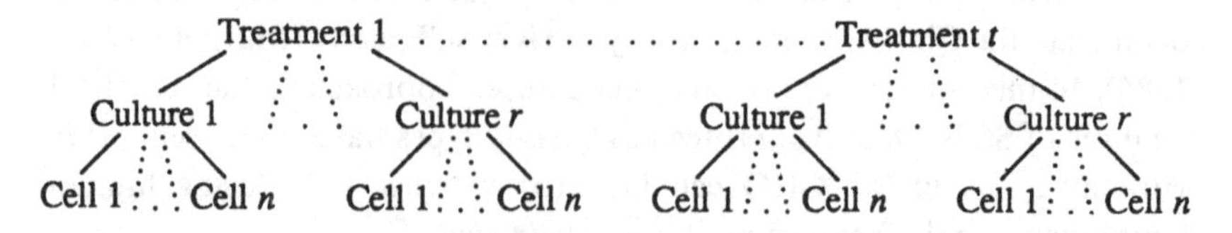

Table 6.2. *SCE counts from foetal cells of mice.*

Dose (mg/kg)	Animal	SCE counts per cell										Total
0	1	3	5	5	6	7	8	10	11	13	14	
		15	17	21	7	9	9	12	12	17	25	226
	2	3	3	4	5	6	6	6	6	6	7	
		7	8	8	9	9	11	12	13	13	19	161
	3	5	6	6	8	8	12	15	17	5	5	
		6	8	9	11	11	13	15	16	13	5	194
	4	4	5	6	6	6	7	7	8	8	9	
		9	9	9	10	12	13	14	16	23	24	205
6.25	5	2	2	3	4	4	4	5	6	8	8	
		8	9	10	10	10	11	6	8	13	20	151
	6	6	6	8	9	3	6	6	6	8	9	
		9	9	10	10	10	11	11	11	12	12	172
	7	2	3	4	5	5	5	5	6	7	8	
		8	9	10	11	14	16	16	16	17	18	185
	8	6	7	9	9	10	10	10	10	12	12	
		12	12	13	13	14	14	15	15	16	17	236
12.50	9	6	6	7	8	10	10	10	11	11	12	
		13	13	13	14	16	16	17	19	20	41	273
	10	4	5	5	7	8	9	9	9	9	10	
		10	10	10	10	11	11	16	21	27	28	229
	11	6	7	8	9	10	10	10	11	11	11	
		11	12	12	14	17	18	18	18	20	23	256
	12	5	8	8	9	9	10	11	12	12	13	
		13	13	13	14	15	16	19	19	21	21	261
25.00	13	9	10	12	13	14	16	16	17	17	17	
		19	20	20	20	20	21	25	38	55	13	392
	14	5	6	7	7	10	11	12	12	12	12	
		13	14	14	14	18	18	19	23	23	37	287
	15	12	12	14	14	17	18	19	19	19	20	
		20	22	22	22	22	23	25	26	28	29	403
	16	4	4	5	5	7	7	8	10	11	12	
		13	14	14	14	19	19	21	23	24	27	261
50.00	17	9	9	12	13	19	20	7	11	18	18	
		21	22	26	28	27	33	30	30	24	44	421
	18	12	15	15	19	20	21	22	22	23	28	
		33	26	33	30	29	33	31	33	33	34	512
	19	14	24	26	33	15	16	22	23	23	30	
		35	28	31	27	28	24	30	38	31	41	539
	20	14	16	18	19	22	26	26	73	24	17	
		20	21	21	21	28	30	30	38	40	28	532

Note: There were five doses (0, 6.25, 12.5, 25 and 50 mg/kg) and four animals per dose. 20 cells were scored for SCE for each animal.

partition of the variation within a data set is made by a hierarchical (or nested) analysis of variance (Snedecor & Cochran, 1980, §13.9). The skeleton form is shown in Box 6.1. The error is estimated by the mean square labelled *B*. If the mean squares *B* and *C* are of a similar magnitude, then any variation between cultures, within treatments, is a simple consequence of the variation between cells: cultures are not an additional source of variation. If this is the case, then error is better estimated from the 'between cells' mean square *C*, or better still, from the mean square derived by pooling 'between cultures' and 'between cells'. These estimates of error are much better than that derived from the 'between cultures' mean square because they are based on many more degrees of freedom.

The statistical test for the presence of culture effects within treatments is made using an *F*-test on the ratio of mean squares B/C. If this is significant then the error for testing treatment effects is estimated by the mean square *B*. If the *F*-test is not significant then the error may be estimated by the mean square obtained by pooling the 'between cultures' and the 'between cells' sums of squares. For convenience, in this latter case we shall say we are using a 'between-cells error' while in the former case we say we are using a 'between-cultures error'.

The choice of appropriate error is important for both the design and analysis of experiments. It is considered desirable that the degrees of freedom of an estimate of error should not be too small, and preferably greater than 10. For an experiment with t treatments, r cultures per treatment and n cells per culture, the degrees of freedom associated with 'between cultures' is $t(r-1)$; care is needed to ensure that this is not small. Whereas the degrees of freedom associated with the estimates of a 'between cells' error is $t(rn-1)$ and will be much greater than 10 for all except very small experiments.

Note that experiments include positive controls but that the results from these are not used directly in the statistical analysis.

6.2.2 Experiments *in vitro* (e.g. SCE in CHO cells or human lymphocytes)

Although some regulatory authorities do not specify a requirement for repeat studies, greater confidence can be attached to a reproducible response. We outline below a two-tier approach consisting of a first experiment designed to assess SCE frequency at a range of doses, followed by a more thorough (in the sense of statistically rigorous) experiment using appropriate dose levels determined from the first experiment. If only a single experiment is carried out, it should use the

Box 6.1. *Hierarchical (or nested) analysis of variance*

Source of variation	Degrees of freedom	Mean square
Between treatments	$t-1$	A
Between cultures within treatments	$t(r-1)$	B
Between cells within cultures	$tr(n-1)$	C
Total	$trn-1$	

Note: t = number of treatments; r = number of replicates per treatment; n = number of cells per replicate.

design of the second experiment, which has greater statistical power.

The minimum requirements for the initial study, according to OECD and EPA recommendations, are that it should comprise a negative (vehicle) control, a positive control and three dose levels of test material; duplicate cultures should be used for positive and negative controls and the dose levels, and the study should be performed in the presence and absence of a supplementary liver enzyme system. (Data from the two metabolic conditions are analysed as separate experiments.) If the between-cultures estimate of error has to be used for testing treatment effects, this experimental design is not very sensitive since the estimate is based on only four degrees of freedom. At least quadruplicates of the negative control are desirable to increase the error degrees of freedom to six. There is also a good case in this exploratory stage to sample well the possible range of response and include five dose levels, each duplicated. This latter set of treatments with a negative control in quadruplicate would give an error estimate based on eight degrees of freedom. If the between-cells estimate of error is the appropriate one then simple duplicates of controls and dose levels may be used.

Three dose levels should be used in the follow-up experiment, spanning the region where a response was found in the initial experiment or, if no response was found, surrounding the highest dose used initially. When the between-cultures estimate of error is appropriate, there should be quadruplicate cultures of the dose levels and the negative control; duplicate cultures of the positive control are adequate. This will yield an error estimate based on 12 degrees of freedom. When the between-cells estimate of error is appropriate, duplicates of controls and dose levels are adequate.

Choice of dose levels is discussed in Section 6.3.5.

6.2.3 Experiments *in vivo*

OECD have not produced guidelines. EPA recommend that an initial assessment should involve three experimental treatments: test material at the maximum tolerated dose or that dose producing some evidence of toxicity, together with concurrent positive and negative controls. There should be five male and five female rodents for each treatment. A single dose is given and a single sampling time used. This type of study uses 30 animals and provides an estimate of error, between animals, with 16 degrees of freedom.

Since animals of different sex are included, sex must be recognised as a factor in the analysis. Assuming t treatments (involving the negative but excluding the positive control) and r replicates of each Treatment × Sex combination, the skeleton analysis of variance is given in Box 6.2. If the Treatment × Sex interaction is significant the two sexes must be considered separately. Our experience with animal experiments is that there are always significant animal effects so we assume in this section that error is estimated from the 'between animals' mean square.

We would recommend that more dose levels are used in an initial study as we recommend for *in vitro* experiments. Thus, *three* dose levels might be used together with negative and positive controls. If three male and three female animals are used for each experimental treatment, the total number of animals involved (30) and the degrees of freedom for error (16) are as before, but the larger number of dose levels would be more informative. Even more dose levels, with reduced replication (subject to a minimum of two males and two females per treatment) might be considered in the initial study, though this depends on the variation between animals.

For the investigation of a dose-response, EPA recommend at least three dose levels with five male and five female animals per treatment. We agree that three dose levels should be used, together with positive and negative controls, but think that, depending on inter-animal variation, consideration should be given to reducing the replication to four or even three animals of each sex.

6.3 PRELIMINARIES TO ANALYSIS

6.3.1 Introduction

There are two major decisions that must be made before analysis. The first is that introduced in Section 6.2.1: whether the between-cultures or the between-cells estimate of error is the appropriate one. The decision must be based on the analysis of experimental data, and preferably on that

Box 6.2. *Analysis of variance to test Treatment × Sex interaction*

Source of variation	Degrees of freedom	Mean square
Treatments	$t-1$	A
Sex	1	B
Treatments × sex	$t-1$	C
Between animals	$2t(r-1)$	D
Total	$2rt-1$	

Note: t = number of treatments; r = number of animals per treatment.

of many similar experiments rather than deciding separately for each experiment. We describe appropriate tests in Section 6.3.3.

The second major decision is what *form* of distribution do the data belong to. A major criterion is whether the distribution is Poisson or not; an appropriate test is described in Section 6.3.3.1.

The situation in which

(a) the SCE counts per cell have a Poisson distribution,
(b) there are no replicate differences,

is worth recognising explicitly. When these conditions hold we say that 'a simple Poisson model' is applicable.

We recommend the analysis of variance of transformed SCE data as a satisfactory and generally applicable method. The two decisions we have already discussed lead to the choice of an appropriate transformation and the form of the analysis of variance table. A procedure for making the decisions is summarised in the flow diagram in Fig. 6.2. The first step is to use a dispersion test (Section 6.3.3.1) to see if the data have a Poisson distribution. If they do, a square root transformation is used. Otherwise, an appropriate transformation must be chosen (Section 6.3.4). The transformed data are used in an analysis of variance (Section 6.3.3.3) to test if there are replicate differences. This decides the form of the final analysis of variance. If there are replicate differences the distribution that matters is that of cultures (animals) and not that of cells. Therefore it must be considered whether the transformation chosen on the basis of the distribution of the individual cell SCE counts is still appropriate, or whether a new transformation is necessary.

For Poisson data, alternative methods without transformation are easily available. Testing replicate effects is discussed in Section 6.3.3.2 and testing linear trend in Section 6.5.1. Margolin *et al.* (1986) give a good general discussion.

6.3.2 Investigation of distribution of number of *SCE*

SCE data consist of counts on cells. A reasonable model to consider is that the counts have a Poisson distribution. An important property of the Poisson distribution is that its variance is equal to its mean. This may be used to test if data follow a Poisson distribution as is shown in the next section.

Margolin *et al.* (1986) described their investigation of SCE counts with *in vitro* tests using Chinese hamster ovary cells according to the protocol of Galloway *et al.* (1985). Their experimental unit was a flask (i.e. a culture in our terminology) and SCE counts were made on 50 cells in each culture. They found that the variability between cells within a culture and the variability between cultures were both consistent with a Poisson model. They noted that the inter-day variability exceeded that of a Poisson model.

Fig. 6.2. Flow diagram for choosing method of analysis.

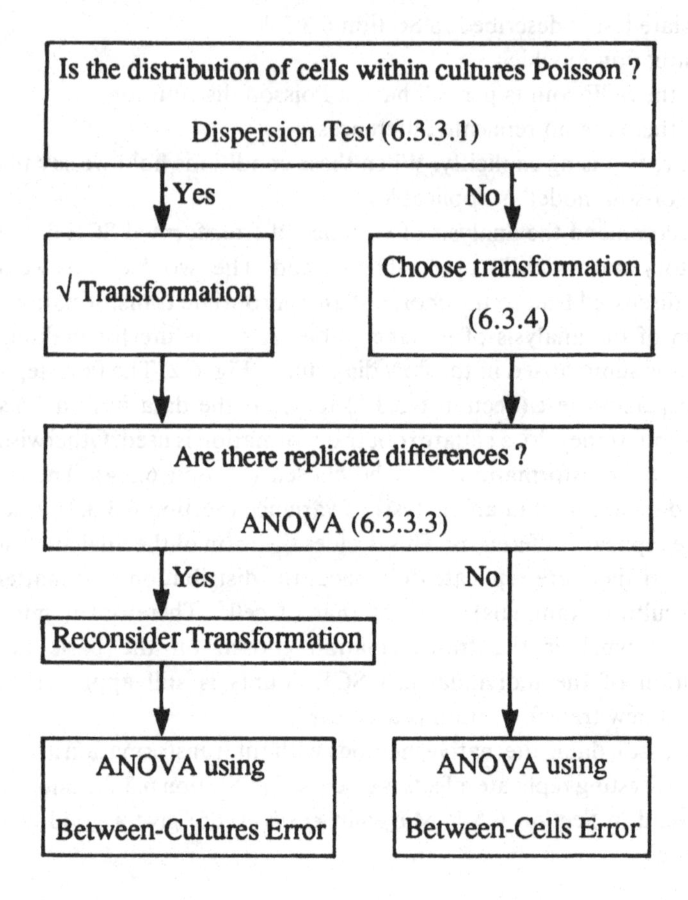

Many workers have found with SCE counts on human lymphocytes that the variability between cells is greater than would be expected on a Poisson model. Wulf *et al.* (1984) found this and made the assumption that the mean count of individual cells itself had a gamma distribution. Their data were consistent with this assumption and they described the distribution of cells within an individual as being a mixed Poisson. Carrano & Moore (1982) also looked at the between-cells distribution of SCE counts in human lymphocytes. They rejected the Poisson as an appropriate model and further concluded that 'no single family of distributions can be used as a model' for all their data sets. Pesch *et al.* (1984) investigated the distribution of SCE counts with the marine worm, *Neanthes arenaceodentata*. They found the variability was higher than would be consistent with a Poisson model and concluded that the variance was proportional to the mean squared rather than to the mean.

Our own limited statistical investigations indicate that SCE counts on CHO cells are Poisson distributed, with no replicate effects, but that counts on lymphocytes and on cells from animal experiments do not have a Poisson distribution.

6.3.3 Testing the observed distributions

6.3.3.1 *Testing distribution of cells within cultures*

To test if the number of SCE for the cells in a culture has a Poisson distribution, a dispersion test (Snedecor & Cochran (1980) §11.4) may be used. An example is shown in Box 6.3 based on the CHO data in Table 6.1. It is based on the property of the Poisson distribution that the variance is equal to the mean. If there are n cells in a culture and the SCE count for the k-th cell is y_k, where $k = 1,2,\ldots,n$, then the dispersion test uses the statistic

$$\chi^2 = \sum_{k=1}^{n} (y_k - \bar{y})^2/\bar{y}$$

where $\bar{y} = \sum_{k=1}^{n} y_k/n$. The statistic equals $(n-1)$ times the ratio of the sample variance to the mean, expressed here as the sum of squares divided by the mean. On the null hypothesis that the values are Poisson, the χ^2 statistic has a chi-squared distribution with $(n-1)$ degrees of freedom. The statistic may be tested for each culture but it is convenient to make a single overall test using the sum of the χ^2 values which, on the null hypothesis, has a chi-squared distribution with degrees of freedom equal to the sum of the degrees of freedom for the individual cultures (see Box 6.3). The number of cells per culture need not be equal.

Box 6.3. *Tests on distribution of counts (Snedecor and Cochran, §11.4)*

1. Testing if distribution of cells within cultures is Poisson (data from Table 6.1). y_k is number of SCE for the k-th cell, where $k = 1, 2, \ldots, n$.

Culture	1	2	3	4	5	6	7	8	Total
$\sum_{k=1}^{n} (y_k - \bar{y})^2$	68.00	43.60	112.10	66.50	60.00	17.60	50.90	58.10	
mean, $\bar{y}$	7.00	6.80	8.30	7.50	6.00	7.80	7.90	8.70	
χ^2	9.71	6.41	13.51	8.87	10.00	2.26	6.44	6.68	63.88
Degrees of freedom	9	9	9	9	9	9	9	9	72

Notes:

$\chi^2 = \sum_{k=1}^{n} (y_k - \bar{y})^2/\bar{y}$. χ^2 values for cultures may be tested individually.
$\chi^2_9(P=0.05) = 16.9$.
Total χ^2 is 63.88 *NS*. $\chi^2_{72}(P=0.05) = 92.8$.
Data not significantly different from a Poisson distribution.

2. Testing existence of replicate effects with Poisson data.

   Variate: T is a replicate total, $\sum_{k=1}^{n} y_k$ (data from Table 6.1). T_j is total for j-th replicate, where $j = 1, 2, \ldots, r$.

Treatment	0	1	2	3	Total
$\sum_{j=1}^{r} (T_j - \bar{T})^2$	2	32	162	32	
Mean, $\bar{T}$	69	79	69	83	
χ^2	0.03	0.41	2.35	0.39	3.18
Degrees of freedom	1	1	1	1	4

Notes:

$\chi^2 = \sum_{j=1}^{r} (T_j - \bar{T})^2/\bar{T}$. Total $\chi^2 = 3.18$ *NS*. $\chi^2_4(P=0.05) = 9.49$.
No evidence of replicate effects since the χ^2 test is not significant.
The conclusion, using both tests, is that the data of Table 6.1 are consistent with a 'simple Poisson model'.

6.3.3.2 *Testing replicate effects with Poisson data*

If the data are consistent with a Poisson model then a test, similar to that in the previous section, may be made to investigate whether replicate cultures of a treatment have the same mean (see Box 6.3). It is assumed, in the first instance, that there are an equal number of cells n per culture and r replicate cultures per treatment.

Let the *total* count for replicate j be T_j and let the mean of the replicate totals for a particular treatment be $\overline{T} = \sum_{j=1}^{r} T_j/r$. The χ^2 statistic is calculated for each treatment and has $(r-1)$ degrees of freedom but, again, it is convenient to test the total χ^2 obtained by summing over the treatments. This is tested as chi-squared with degrees of freedom equal to the sum of the degrees of freedom of the individual treatments. Rejection of the null hypothesis is evidence that there are differences between replicates so that what we have called the 'simple Poisson model' does not hold. Non-rejection of the null hypothesis shows that the data are consistent with there being no replicate effects but the evidence is strong only if data from many experiments have been tested.

If there are a varying number of replicates from treatment to treatment (e.g. r_i replicates of Treatment i), the only difference in the test is that the degrees of freedom of the χ^2 corresponding to Treatment i becomes (r_i-1). If, in addition, the number of cells per culture varies so that there are n_j cells in culture j of a particular treatment, the chi-squared statistic for that treatment will be $\sum_{j=1}^{r_i} (T_j - n_j\bar{y})^2/n_j\bar{y}$, where $\bar{y}$ is the mean per cell for the treatment, i.e. $\bar{y} = \sum_{j=1}^{r_i} T_j/\sum_{j=1}^{r_i} n_j$. The corresponding degrees of freedom for Treatment i will be (r_i-1). Again, the test is best made with the total chi-squared statistic.

6.3.3.3 *A general method of testing replicate effects*

A method that may be used whether the data have a Poisson distribution or not is based on the hierarchical (or nested) analysis of variance that has already been described in Section 6.2.1. But usually, with SCE data, in order that the assumptions of the analysis of variance hold it

Box 6.4. *General method of testing existence of replicate culture effects (Snedecor & Cochran, 1980, §13.9)*
The square-root transformation is appropriate for the data of Table 6.1.
Variate: $z = \sqrt{y}$, where y is number of SCE per cell.

Source of variation	Degrees of freedom	Sum of squares	Mean square	F
Treatments	3	1.0344	0.3448	
Cultures in treatments	4	0.9624	0.2406	0.96
Cells in cultures	72	18.0274	0.2504	
Total	79	20.0242		

Note:
$F = 0.2406/0.2504 = 0.96$ *NS*. $F_{4,72}(P = 0.05) = 2.50$.
No evidence of replicate culture effects.

will be necessary to transform the data. The choice of a transformation is discussed in the next section (6.3.4); we shall assume here that an appropriate transformation of the data has been made.

We give an example (Box 6.4) based again on the CHO data of Table 6.1. This allows a direct comparison with the method of the previous section. The square-root transformation is appropriate. The test is made with the *F*-statistic which is the ratio of the 'cultures in treatments' mean square to the 'cells in cultures' mean square. If this exceeds the tabular value at the chosen significance level then the conclusion is that there are replicate effects. Otherwise we say there is no evidence of replicate effects as is the case in the example.

6.3.4 Choice of a transformation

6.3.4.1 *Introduction*

We recommend an approach using the analysis of variance but the validity of the method depends on the data being approximately normally distributed with constant variance. Data which do not satisfy the assumptions are commonly transformed so that the assumptions are better satisfied. In the present context, the most important aim of transformation is to ensure that the data analysed are of (approximately) constant variance.

6.3.4.2 *Poisson distribution*

If observations y have a Poisson distribution then the square roots $\sqrt{y}$ have approximately constant variance, and may be analysed by conventional methods (Snedecor & Cochran, 1980, §15.11). The values of

SCE per cell for each experimental unit are transformed. The residual mean square (e.g. the 'between cells' mean square) in an analysis of variance should approximately equal 0.25. Wulf *et al.* (1984) used an almost identical transformation '$\sqrt{y}+\sqrt{(y+1)}$' which has the convenience of having an expected residual mean square of unity but back-transformation is awkward. This transformation and $\sqrt{(y+1)}$ are also better than $\sqrt{y}$ when some of the counts are small.

6.3.4.3 *Other distributions*

For data which do not have a Poisson distribution, an appropriate transformation to stabilise the variance must be determined. One method (Healy, 1968) is to find the relation between the standard deviation s and the mean m by plotting log s against log m and calculating the fitted line $\log s = a + b \log m$. A plot derived from the animal data of Table 6.2 has been made in Fig. 6.3. The mean and standard deviation of the SCE values were calculated for each of the 20 animals. The usual least squares fit gives a slope $b = 0.727$ with standard error 0.17. The recommendation is then to transform from the original observations y to y^{1-b} (or to $\log y$ if $b = 1$). It is not necessary to use the precise index $(1 - b)$; usually the nearest integer or half-integer is chosen. For example, for values of b between 0.25 and 0.75 a square root transformation would be chosen, whereas for values of b between 0.75 and 1.25 the choice would be a logarithmic transformation. For the example data, the value of $b = 0.727$ is on the borderline of indicating a square root or a logarithmic transformation. This type of investigation is best done with many sets of data from similar experiments and an overall choice made; use of more than one transformation makes comparison of experiments difficult. But since we must choose a transformation for our example, to follow through the analysis later, we choose a square root transformation.

6.3.5 Scale for dose

The chosen dose levels may be denoted by $d_2, d_3, \ldots$ and that of the negative control, or zero dose, by d_1. The results are more easy to interpret if the plot of response against dose is a straight line. Therefore another scale (e.g. log d) may seem more appropriate for the dose. We introduce a new variable x, which may be a transformed value of d. (The variable x has been called the dose metameter. This is useful when a distinction is being made between d and x, but when there is no ambiguity we shall refer to either as 'dose'.) In the discussion of analysis we shall use x as the measure of dose; it may be identical with d or be some function of d such as a linear transformation of the logarithm of d.

Analysis is much simplified if the dose metameter levels, including the negative control (i.e. $x_1, x_2, x_3, \ldots x_t$) are equally spaced on an arithmetic scale. This will be so if the control and dose levels (i.e. $d_1, d_2, d_3, \ldots, d_t$) are chosen at equal intervals on an arithmetic scale (e.g. 0, 10, 20, ...) and x is equal to d or is a linear function of d. Frequently, the doses $d_2, d_3, d_4, \ldots d_t$ are chosen to be at equal intervals on a logarithmic scale (e.g. 1, 3, 9, ...). Then x may be chosen to equal log d or be a linear function of log d. There is a problem, however, in deciding how to incorporate the negative control or zero dose treatment, because the logarithm of zero (log 0) is an undefined quantity. Sister chromatid exchanges are present even at zero dose, so one

Fig. 6.3. Plot of log(standard deviation) against log(mean). Each of the 20 points is derived from the cell counts of SCE for each animal. The data are in Table 6.2.

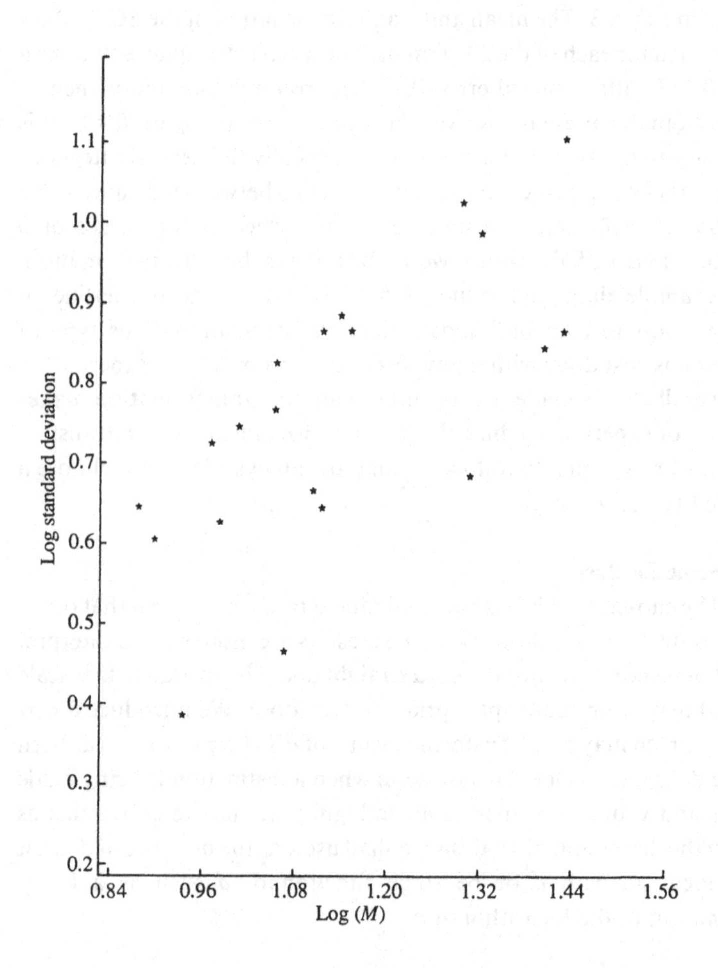

may think of a causative factor of a certain strength operating at zero dose to which is added the test doses at various strengths. This suggests that the location of the zero dose on the x scale should not be too far below the position of the lowest non-zero dose and certainly not close to 'minus infinity' as a formal consideration of log 0 might suggest. We recommend the rule of Margolin *et al.* (1986) that 'x_1 be set below x_2 by an amount that is equal to the average spacing between consecutive values of x', i.e. assuming doses $x_2, \ldots, x_t$, the control dose is given by

$$x_1 = x_2 - [(x_t - x_2)/(t-2)].$$

In the example above (with $d_2 = 1$, $d_3 = 3$, $d_4 = 9$) the logarithms of the values (i.e. 0, log 3, 2 log 3) may be linearly transformed ('divide by log 3 and add 1') to give the values $x_2 = 1$, $x_3 = 2$, $x_4 = 3$. Using the rule above gives $x_1 = 0$, so the four remaining values of x are equally spaced, and take convenient values.

The choice of equally spaced x values is not always possible since it may conflict with the constraints of sampling near the toxic limit and the wish to include several doses in the toxic range.

6.3.6 Unequal number of chromosomes per cell

The aim should be to record the complete set of chromosomes in each cell. This may not always be possible because of chromosome damage, or aneuploidy, so that an equal number of chromosomes is not recorded in every cell. Before beginning analysis, the SCE counts of cells should be adjusted, where necessary, to what they would be if a normal complete set of chromosomes had been counted.

6.4 ANALYSIS OF TRANSFORMED DATA

6.4.1 Introduction

Having chosen an appropriate transformation (Section 6.3.4), an analysis of variance (*ANOVA*) is carried out. From previous tests (Section 6.3.3) it must be decided whether a between-cells or a between-cultures estimate of error is appropriate. The major difference this makes is whether the basic observations used in the analysis are cell values or culture (animal) values. Descriptions of the two analyses, mainly through the use of examples, are given in Sections 6.4.2 and 6.4.3.

A table of means gives a convenient summary of the experimental results. Significance tests can be done only on the transformed data but it is informative to quote the mean values obtained by back-transformation on a cell and a chromosome basis. A standard error for the differences of the means (of the transformed data) should be quoted. One method of testing the treatment effect is to compare each of the dose means with the control using a t-test. But a more sensitive method, usually, is to test if there is a significant linear trend. Since, if there is a positive response to the test substance, one would expect the number of SCE to increase with dose over all, or much of, the range of doses. A one-tail test is appropriate. The test is illustrated in the examples in Sections 6.4.2 and 6.4.3.

Some investigation of the form of the response curve may be made. The treatments sum of squares may be partitioned into a 'linear' sum of squares and a 'deviations from linearity' sum of squares. If the latter is significant, it is evidence that there is non-linearity of response. For example, the change in number of SCE with increasing dose may be positive at low levels of dose but negative at the high levels. The examples illustrate how the test is made. If appropriate computer programs are available, a complete partition of the treatments sum of squares into single degrees of freedom corresponding to polynomials of various degrees may be made. This also is illustrated.

When animals of both sexes are used in an experiment, a test should be made of whether male and female animals are responding to treatments in a similar way (Section 6.4.4).

6.4.2 *ANOVA* using between-cells error

The analysis will be shown by an example, using the CHO data of Table 6.1. A preliminary analysis of these data (in Section 6.3.3) showed that the 'simple Poisson model' is applicable. Therefore the analysis of variance will be carried out on the square roots of the cell values and the error variance will be estimated by the between-cells within-treatments mean square.

The test doses of 1, 3.3 and 10μg/ml were chosen to be equally spaced on a logarithmic scale. As in the example of Section 6.3.5, these may be coded 1, 2, 3 and the control treatment coded 0.

A one-way classification analysis of variance (Box 6.5), as described in Snedecor and Cochran (1980, §12.3) is carried out on the 80 values obtained by taking square roots of the cell values in Table 6.1. There are 4 treatments and 20 replications of each treatment. The estimate of error is given by the 'cells within treatments' or 'between cells' mean square 0.2499. For Poisson data, when analysed by the square root transformation, the expected value of this mean square is 0.25. There is therefore very good agreement.

The linear trend and its standard error are estimated as shown in Box 6.5. The one-tail t test does not give a significant result. The conclusion is that the chemical is not having an effect on the SCE count.

Part (a) of Box 6.6 shows how the treatments sum of squares in the

Box 6.5. ANOVA *using square-root transformation and between-cells error. One-way* ANOVA *(Snedecor & Cochran, 1980, §12.3)*
Variate: $z=\sqrt{y}$, where y is the SCE count of a cell.
Data from Table 6.1.

Source of variation	Degrees of freedom	Sum of squares	Mean square	F
Treatments	3	1.0344	0.3448	1.38
Cells within treatments	76	18.9898	0.2499 ($=s^2$)	
Total	79	20.0242		

Table of means

Dose (μg/ml)	0	1	3.3	10	*SE* of difference
Coded log dose, x_i	0	1	2	3	of means $\sqrt{(2s^2/nr)}$
Transformed data					
Mean $\bar{z}_i$, per cell	2.586	2.752	2.583	2.850	0.158
Back-transformed data					
Mean, per cell	6.69	7.57	6.67	8.12	
Mean, per chromosome (CHO cells, 2n = 20)	0.334	0.379	0.334	0.406	

Linear trend

$$b=\frac{\sum_{i=1}^{t}(x_i-\bar{x})\bar{z}_i}{\sum_{i=1}^{t}(x_i-\bar{x})^2}$$

$$=\frac{(-1.5)\times 2.586+(-0.5)\times 2.752+0.5\times 2.583+1.5\times 2.850}{(-1.5)^2+(-0.5)^2+(0.5)^2+(1.5)^2}$$

$$=\frac{0.3115}{5}=0.0623$$

SE of $b=\sqrt{[s^2/nr\sum(x_i-\bar{x})^2]}=\sqrt{[0.2499/(10\times 2\times 5)]}=\sqrt{(0.002\,499)}$ $=0.049\,99$.

$t=0.0623/0.0499=1.25(t_{76}(P=0.05)=1.67$, one-tail)

There is not a significant linear trend.

analysis of variance may be partitioned into 'linear' and 'deviations from linearity' components. Since the linear trend was not significant, there is no evidence of a straight line response. Moreover, since 'deviations from linearity' are not significant (as tested by the F value of 1.29 shown in Box 6.6) there is no evidence of a curved response. Part (b) shows the complete partition of the treatments sum of squares into linear, quadratic and cubic

Box 6.6. *Testing form of response curve (between-cells* ANOVA*).* Variate: $z=\sqrt{y}$, where y is the SCE count per cell (Data from Table 6.1)

(a) partial decomposition of treatments sum of squares

Source of variation	Degrees of freedom	Sum of squares	Mean square	*F*
Linear	1	0.3881		
Deviations from linearity	2	0.6463	0.3232	1.29
Treatments	3	1.0344		
Between cells	76	18.9898	0.2499	
Total	79	20.0243		

$$\text{Linear sum of squares} = nr\frac{\left[\sum_{i=1}^{t}(x_i-\bar{x})\bar{z}_i\right]^2}{\sum_{i=1}^{t}(x_i-\bar{x})^2} = (10\times 2)\frac{(0.3115)^2}{5} = 0.3881$$

'Deviations from linearity' SS = Treatments SS – Linear SS. For testing deviations from linearity, calculated $F=1.29$ $F_{2,76}(P=0.05)=3.117$ not significant.

(b) complete decomposition of treatments sum of squares

Source of variation	Degrees of freedom	Sum of squares	Mean square	*F*
Linear	1	0.3881	0.3881	
Quadratic	1	0.0515	0.0515	0.21
Cubic	1	0.5948	0.5948	2.38
Treatments	3	1.0344		
Between cells	76	18.9898	0.2499	
Total	79	20.0243		

For testing quadratic and cubic effects $F_{1,76}$ $(P=0.05)=3.97$. Comparison with calculated F values in *ANOVA* table shows that neither is significant.

components using orthogonal polynomials. This is possible provided a computer program or statistical expertise is available. No component is significant.

6.4.3 *ANOVA* using between-cultures (animals) error

This analysis will be illustrated by using the animal data of Table 6.2. The dispersion test of Section 6.3.3.1 showed that these data do not

Box 6.7. *Testing existence of replicate animal effect.*
Variate: $z=\sqrt{y}$ is the SCE count per cell (Data from Table 6.2)

Source of variation	Degrees of freedom	Sum of squares	Mean square	F
Treatments	4	205.93	51.48	
Animals in treatments	15	29.17	1.94	3.07
Cells in animals	380	241.03	0.6343	
Total	399	476.13		

For testing existence of replicate animal effects, calculated $F=3.07$. ($F_{15,380}$ $(P=0.05)=1.69$.) Hence there is evidence of replicate effects.

Table 6.3. *Variate is sum of square roots of the 20 SCE scores for each animal given in Table 6.2.*

Dose (mg/kg)					Total
0	65.22	55.24	60.96	62.20	243.62
6.25	52.90	58.00	58.42	68.16	237.48
12.50	71.88	65.46	70.46	71.22	279.02
25.00	86.36	73.66	89.14	69.56	318.72
50.00	89.44	100.18	102.90	101.04	393.56

Note: Values stated for each of the four animals for each of the five doses.

have a Poisson distribution. The data were used to illustrate how to choose a transformation for non-Poisson data (Section 6.3.4.3) and it was decided to use a square-root transformation. The final stage in the preliminary analysis is to decide if there are replicate effects, using the hierarchical analysis of variance method of Section 6.3.3.3. The analysis of variance is shown in Box 6.7. There is strong evidence of the existence of replicate (animal) effects. Since the animal to animal variation is large the analysis will be carried out on the individual animal total values (which are the sum of the transformed cell values). These data are shown in Table 6.3.

By a similar argument to the previous section, and Section 6.3.5, the treatments are coded 0, 1, 2, 3, 4. A one-way classification analysis of variance with five treatments and four replications of each treatment is carried out (Box 6.8). Note that the values in the first two lines of the analysis of variance table are similar to those in the hierarchical analysis table of Box 6.7, except that the former values are 20 times the latter. (In general, the multiplier is n, the number of cells per culture or per animal.) The error is estimated by the 'between animals' mean square, 38.89.

Box 6.8. ANOVA *using between animals (cultures) error. One-way* ANOVA *(Snedecor & Cochran, 1980, §12.3)*

Variate $z = \sum_{k=1}^{n} \sqrt{y_k}$, where y_k is *SCE* count on k-th cell for an animal

(Data in Table 6.3)

Source of variation	Degrees of freedom	Sum of squares	Mean square	F
Treatments	4	4119.79	1029.95	26.48
Between animals	15	583.34	38.89($=s^2$)	
Total	19	4703.13		

Table of means

Dose (mg/kg)	0	6.25	12.5	25	50	*SE* of difference of means
Coded log dose, x	0	1	2	3	4	
Transformed data						
Mean $\bar{z}$, per animal	60.90	59.37	69.76	79.68	98.39	4.41 $\sqrt{(2s^2/r)}$
Mean, per cell	3.04	2.97	3.49	3.98	4.92	0.22 $\sqrt{(2s^2/rn^2)}$
Back-transformed data						
Mean, per cell	9.27	8.81	12.16	15.87	24.20	
Mean, per chromosome (Mouse cells, $2n=40$).	0.232	0.220	0.304	0.397	0.605	

Linear trend

$$b = \frac{\sum_{i=1}^{t} (x_i - \bar{x})\bar{z}_i}{\sum_{i=1}^{t} (x_i - \bar{x})^2} = \frac{(-2)\times 60.905 + (-1)\times 59.370 + 79.680 + 2\times 98.390}{(-2)^2 + (-1)^2 + 1^2 + 2^2}$$

$$= \frac{95.28}{10} = 9.528$$

$$SE \text{ of } b = \sqrt{\left[s^2/r \sum_{i=1}^{t} (x_i - \bar{x})^2\right]} = \sqrt{[38.89/(4\times 10)]} = \sqrt{0.97225)} = 0.986$$

$t = 9.528/0.986 = 9.66$. t_{15} ($P = 0.001$, one tail) $= 3.73$.
There is a significant linear trend.

In the table of means, it should be noted that it is the mean per cell (and not *per animal*) that is back-transformed, by squaring, because it was the *cell* values in the original data that were transformed.

The linear trend is tested and found to be highly significant. There is therefore very strong evidence of an effect of the chemical on the SCE count. Investigation of the response curve (Box 6.9) shows a significant effect of 'deviations from linearity' in part (a), indicating a curvature effect

Box 6.9. *Testing form of response curve (between-cells* ANOVA*).*

Variate: $z=\sum_{k=1}^{n}\sqrt{y_k}$, *where* y_k is SCE count on k-th cell for an animal

(Data from Table 6.3)

(a) partial decomposition of treatments sum of squares

Source of variation	Degrees of freedom	Sum of squares	Mean square	F
Linear	1	3631.31	3631.31	
Deviations from linearity	3	488.48	162.83	4.19
Treatments	4	4119.79		
Between animals	15	583.34	38.89	
Total	19	4703.13		

$$\text{Linear sum of squares} = r\frac{\left[\sum_{i=1}^{t}(x_i-\bar{x})\bar{z}_i\right]^2}{\sum_{i=1}^{t}(x_i-\bar{x})^2} = 4\frac{(95.28)^2}{10} = 3631.31$$

'Deviations from linearity' SS = Treatments SS – Linear SS. For testing deviations from linearity, $F=4.20$. $F_{3,15}$ $(P=0.05)=3.29$. Hence significant.

(b) complete decomposition of treatments sum of squares

Source of variation	Degrees of freedom	Sum of squares	Mean square	F
Linear	1	3631.31	3631.31	
Quadratic	1	457.83	457.83	11.77
Cubic	1	3.93	3.93	0.10
Quartic	1	26.72	26.72	0.69
Treatments	4	4119.79		
Between animals	15	583.34	38.89	
Total	19	4703.13		

For testing quadratic, cubic and quartic effects $F_{1,15}$ $(P=0.05)=4.54$. Comparison with calculated F values in *ANOVA* table shows that the quadratic effect is significant.

as well as the linear trend. Part (b) shows it to be a significant quadratic response or parabolic response. A graph, Fig. 6.4, illustrates the situation. A least-squares straight line drawn through the points would give a good approximate summary of the response. But because the points corresponding to the extreme doses would be above the line, while the three points corresponding to the central doses would be below the line, a parabola drawn through the points would provide a better description.

6.4.4 Analysis with animals of both sexes

If animals of both sexes are used in an experiment, the analysis should take account of the factorial structure with the factors sex and treatments. The data of Table 6.3 will be used again but of the four animals for each treatment, which are in random order, 1 and 2 will be assumed male, 3 and 4 will be assumed female. The analysis is shown in Box 6.10. A two-way table of treatments × sex totals is constructed and used to calculate the analysis of variance.

The important test is that of the treatments × sex interaction. If the

Fig. 6.4. Plot of mean of the square-rooted cell counts of SCE against coded log dose. The basic data are in Table 6.2.

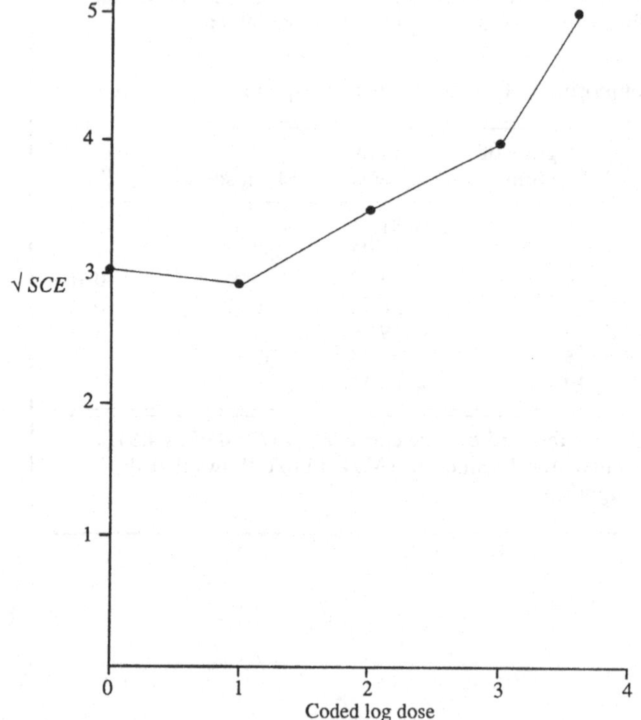

Box 6.10. *Testing treatments × Sex interaction.*

Variate: $z = \sum_{k=1}^{n} \sqrt{(y_k)}$, where y_k is SCE count on k-th cell for an animal
(Data in Table 6.3)

Treatments × Sex table (totals of two animals)

	Sex		
Dose	Male	Female	Total
0	120.46	123.16	243.62
6.25	110.90	126.58	237.48
12.50	137.34	141.68	279.02
25.00	160.02	158.70	318.72
50.00	189.62	203.94	393.56
Total	718.34	754.06	1472.40

Source of variation	Degrees of freedom	Sum of squares	Mean square	F
Treatments	4	4119.79		
Sex	1	63.80		
Treatments × Sex	4	55.90	13.975	0.30
Between animals	10	463.64	46.364	
Total	19	4703.13		

Correction Factor $(CF) = 1472.40^2/20 = 108398.0880$.
Sex $SS = (718.34^2 + 754.06^2)/10 - CF = 63.80$.
Treatments $SS = (243.62^2 + \ldots + 393.56^2)/4 - CF = 4119.79$
Treatments × Sex $SS = (120.46^2 + 123.16^2 + \ldots + 189.62^2 + 203.94^2)/2$
$- CF -$ Sex $SS -$ Treatments $SS = 55.90$.
Testing Treatments × Sex, $F = 13.975/46.36 = 0.30$ $F_{4,10}$ $(p = 0.05) = 3.48$.
No evidence of interaction.

interaction is significant, this implies that the male and female animals are reacting in different ways. The treatment totals for male and female animals would then have to be analysed separately. Not surprisingly, in the example the interaction is not significant. This being so, testing proceeds as in the example of the previous section, except that the appropriate estimate of error is that found from the analysis of variance in Box 6.10.

6.5 OTHER METHODS OF ANALYSIS

6.5.1 Testing linear trend when simple Poisson model applicable

When the requirements for the simple Poisson model (Section 6.3.1) are satisifed, a test of linear trend may be made without transforming the data. The method is shown in Box 6.11. An estimate of the linear trend of treatment totals (where the summation is over cultures and cells) is made. Assuming the null hypothesis of no treatment difference (i.e. that the trend is zero) it is possible to calculate a standard error for the estimate of trend. The ratio of trend estimate to standard error, denoted by z, is approximately normal and may be tested by comparison with normal tables. Again, a one-tail test should be used.

6.5.2 Generalised linear model

A more flexible and sophisticated method than that of transformation, is to use what is called a generalised linear model. But to do this the reader will need access to the computer package GLIM and probably need the advice of a statistician. It is not possible, in a limited space, to describe the approach. We restrict ourselves to offering limited guidance on some of the necessary decisions.

Preliminary analyses, as described in Section 6.3, will still be necessary to decide whether (i) the data have a Poisson distribution, (ii) there are culture replicate differences. The second decision determines whether the cell SCE counts are used as the basic data in an analysis (which is the case when there are no culture replicate differences) or whether the culture replicate total counts are used.

The distribution of the data being analysed must be specified using a limited number of possibilities. Choosing the Poisson distribution is likely to be a generally satisfactory approximation even if the data have failed the Poisson tests in the preliminary analysis. For non-Poisson data a variance parameter will have to be estimated using the results of the analysis. A 'link' function must be specified. Either the 'identity link', or the 'log link' is likely to be satisfactory.

6.5.3 Non-parametric methods

Some workers use nonparametric methods because the raw SCE data do not usually satisfy the assumptions required for parametric tests. We prefer to transform the data so that the assumptions are approximately satisfied, because the analysis of variance, and associated methods, are very flexible techniques and are widely understood. A limitation on non-parametric methods is that techniques for deriving confidence intervals are still not well developed.

6.6 SIZE OF EXPERIMENTS

6.6.1 Introduction

If the experimenter has (1) information on the variability of his experimental material, and is willing to say, (2) what size difference he

Box 6.11. *Testing linear trend when simple Poisson model applicable.* (Data from Table 6.1).

Let S_i be the total SCE count for Treatment i, ($i=1, 2, \ldots, t$), summing over replicates and cells.

Dose (μg/ml)	0	1	3.3	10
Coded log dose, x_i	0	1	2	3
Total SCE count, S_i (20 cells)	138	158	138	166

Mean $\overline{S}$, of total counts $= \sum_{i=1}^{t} S_i/t = 600/4 = 150$.

$$\text{Estimate of linear trend of treatment totals} = \frac{\sum_{i=1}^{t} (x_i - \overline{x}) S_i}{\sum_{i=1}^{t} (x_i - \overline{x})^2}$$

$$= \frac{(-1.5)\times 138 + (-0.5)\times 158 + (0.5)\times 138 + (1.5)\times 166}{(-1.5)^2 + (-0.5)^2 + (0.5)^2 + (1.5)^2} = \frac{32}{5} = 6.40$$

Standard error of trend, on null hypothesis of no treatment difference
$= \sqrt{[\overline{S}/\sum(x_i - \overline{x})^2]} = \sqrt{(150/5)} = \sqrt{30} = 5.48$

Test statistic $z = 6.40/5.48 = 1.17$ (Normal z ($P=0.05$) $= 1.64$, one-tail).
No evidence of linear trend in treatment totals.

wishes to pick out and, (3) the probability with which he wishes to detect that difference if it does exist, then it is possible to calculate the minimum number of replications required. The basis of the calculations is discussed in Snedecor & Cochran (1980, §6.14) and in Cochran & Cox (1957, §2.2).

We assume that two means are being compared of which one is a control mean and the other is a treatment mean. The size of difference it is required to pick out must be stated in terms of the scale that is being used. Suppose the control mean is 10 and we wish to detect an increase to 12, which is an increase of $(12-10)/10 = 1/5$ or 20%. In the square root scale the increase is $(\sqrt{12} - \sqrt{10})/\sqrt{10} = 0.095$ or 9.5%, and this is the value that would be used in the formulae in the books referred to above. In fact, if the difference in the raw data is d% (assumed to be in the range 0–50%) then a satisfactory approximation to the difference in the square root scale is $(d/2)$%.

The calculations give the number of replicates required for each of the control and the other treatments. If more treatments are being used, and the trend is being tested, then the stated number of replicates must be used for *each* of three treatments (including the control) but may be reduced by 10% for each treatment, if four treatments are used, and by 20% if five treatments are used.

The probabilities of detecting a difference are usually chosen to be 80% or 90%. To use higher probabilities (e.g. 95% and 99%) gives a very high number of replicates.

6.6.2 Number of cells when simple Poisson model holds

The requirements for the simple Poisson model are stated in Section 6.3.1; CHO data appear to satisfy them. We assume a square root transformation will be used, and that the test will be a 5% one-tail t-test.

Since there are no replicate differences, we calculate the number of cells ($=rn$, with r replicates and n cells per replicate) required for each treatment. The appropriate formula is a slight modification of (6.14.4) in Snedecor & Cochran (1980) and is

$$rn = 2(Z_{2\alpha} + Z_\beta)^2 \sigma^2 / \delta^2.$$

For a 5% one-tail test, $Z_{2\alpha} = 1.645$. Choosing a 90% probability of detecting the difference δ makes $Z_\beta = 1.282$. For Poisson data in a square root scale, the standard deviation equals 0.5. But the σ in the formula is often most conveniently expressed as a proportion of the mean, which in the square root scale is approximately $\sqrt{m}$, where m is the mean SCE count per cell in the original data. Hence we take $\sigma = 0.5/\sqrt{m}$. Finally, if d is the proportional difference from the control mean m in the original data then, in the transformed data, $\delta = d/2$ approximately. Hence a convenient formula for use is

$$rn = 17/(md^2).$$

For example, if we wish to detect a 10% increase in the observed SCE numbers, and the control mean value is 6.9 as in Table 6.1, then $rn = 17/(6.9 \times 0.1 \times 0.1) = 246$. The number of culture replicates is not fixed by this calculation but we might choose to use five cultures and count 50 cells from each. Note that we round off the number of cells to a convenient value; the calculation does not give precise values.

If the probability of detection is dropped to 80%, the only change is that $Z_\beta = 0.842$. The formula becomes

$$rn = 12/(md^2).$$

6.6.3 Number of cells when no culture replicate differences but data not Poisson

The formula to determine the number of cells is again based on that given at the beginning of Section 6.6.2, but in this case σ^2 is not known theoretically as it is for Poisson data, so it must be estimated from previous analyses.

Choosing a 90% probability of detecting the difference δ, and using a 5% one-tail t-test, gives $Z_{2\alpha}=1.645$ and $Z_\beta=1.282$. The formula becomes

$$rn=17\sigma^2/\delta^2.$$

As an example, an estimate of the variance between cells for the data of Table 6.1 may be obtained from Box 6.5. The estimate equals 0.2499 and is obtained from the analysis of variance table. Since the control mean equals 2.586, then σ, expressed as a proportion of the mean, may be taken to be $\sqrt{(0.2499)}/2.586=0.19$ or 19%. Suppose we wish to detect differences of 10% in the original scale, which approximately equals differences of 5% in the square root scale; thus we take $\delta=5\%$ or 0.05. Substituting in the formula, $rn=17(0.19/0.05)^2=245$. This gives a similar answer to the example in Section 6.6.2, as it should since the data are in fact Poisson distributed. Since we have used the observed variance rather than the theoretical variance, we do not obtain an identical answer.

6.6.4 Number of replicates when there are culture replicate differences

The case we now consider is the likely one when animals are used. Using a formula similar to those used in the two previous sections becomes more difficult because the normal (Z) values in the formulae must be replaced by t values. This is a consequence of the number of animals (cultures) being much smaller than the total number of cells. Moreover, the t values depend on the number of animals and an equation must be solved repeatedly until a single value emerges. Most readers will find it more convenient to use the tables in Cochran & Cox (1957, §2.2); we give an example using the animal data of Table 6.3.

Again we choose to consider a 5% one tail test and a probability of 90% of detecting it. To enter the 'Cochran & Cox' table we need values of σ and δ both expressed as a percentage of the mean. From Box 6.8 the between animals variance is estimated by 38.89 and the control mean is 60.9. Hence σ may be estimated by $\sqrt{(38.89)}/60.9=0.102$ or 10.2%. If we wish to detect a difference of 10% of the control mean value in the original data this corresponds to approximately 5% in the square root scale. Hence $\delta=0.05$ or 5%. Using these values in the table shows that the number of replicates required exceeds 50. If the difference it is required to detect is increased to 20% in the original scale, so that $\delta = 10\%$, then the number of replicates required equals 18.

6.7 CONCLUDING REMARKS

6.7.1 Methods of analysis

We recommend that SCE data from *in vitro* and *in vivo* experiments should be analysed by analysis of variance. For each type of experimental data an appropriate transformation should be determined. In our experience, and that of others (Margolin *et al.*, 1986) SCE counts obtained from CHO cells follow a Poisson distribution so a square root transformation can be used. The appropriate transformation for other systems will need to be empirically determined. It is necessary to test for replicate differences (between cultures or animals) to determine the appropriate form of the analysis of variance. The treatment effect can be tested by a linear trend test.

6.7.2 Interpretation of positive responses

The aim of statistical analysis of data is to help decide whether a test chemical is genotoxic. The experimenter should be aware of the possibility that small absolute increases in SCE may be statistically significant and that such results should be considered in the light of artefactual increases in SCE rates. For example, it has been shown that varying the sampling time in CHO cells (Ockey, 1980) or human lymphocytes (Scott, 1982) can produce large fluctuations in SCE frequencies. Perturbations of osmolality or pH of the culture medium may also lead to artefactual results in *in vitro* experiments.

6.7.3 Experimental design

Adequate replication should be included to give sufficient degrees of freedom for sensitive statistical analysis. Historical control data should be analysed to determine variability between cultures (animals) in a particular assay system and this information used to determine the optimal size of experiments. The choice of constant log-intervals or arithmetic increases in the dose levels will simplify statistical analysis. However, this choice may conflict with scientific decisions based on constraints of sampling near the toxic limit and the wish to include several doses in the toxic range.

6.8 REFERENCES

Carrano, A.V. & Moore, D.H.II (1982). The rationale and methodology for quantifying sister chromatid exchange in humans. In *Mutagenicity: New Horizons in Genetic Toxicology*. Ed. J. Heddle, Academic Press. pp. 267–304.

Cochran, W.G. & Cox, G.M. (1957). *Experimental Designs*, 2nd edn. Wiley, New York.

EPA (1983). New and revised health effects test guidelines. EPA Report No. 560/6–83–001.

Galloway, S.M., Bloom, A.D., Resnick, M., Margolin, B.H., Nakamura, F., Archer, P. & Zeiger, E. (1985). Development of a standard protocol for *in vitro* cytogenetic testing with Chinese hamster ovary cells; Comparison of results for 22 compounds in two laboratories. *Environmental Mutagenesis*, **7**, 1–51.

Healy, M.J.R. (1968). The disciplining of medical data. *British Medical Bulletin* **24**, 210–14.

Margolin, B.H., Resnick, M.A., Rimpo. J.Y., Archer, P., Galloway, S.M., Bloom, A.D. & Zeiger, E. (1986). Statistical analyses for *in vitro* cytogenetic assays using Chinese hamster ovary cells. *Environmental Mutagenesis*, **8**, 183–204.

Ockey, C.H. (1980). Differences between 'spontaneous' and induced sister chromatid exchanges with fixation time and their chromosome localisation. *Cytogenetics and Cell Genetics*, **26**, 223–235.

OECD (1986). Guidelines for testing of chemicals. No 479. Genetic toxicology: *in vitro* sister chromatid exchange assay in mammalian cells.

Perry, P., Henderson, L. & Kirkland, D. (1984). Sister chromatid exchange in cultural cells, in: *UKEMS Sub-Committee on Guidelines for Mutagenicity Testing*, Report, Part II, Supplementary Tests, Ed. B.J. Dean, United Kingdom Environmental Mutagen Society, Swansea, pp. 89–121.

Pesch, G., Heltshe, J. & Meuller, C. (1984). A statistical analysis of *Neanthes arenaceodentata.* sister chromatid exchange data. In *Sister Chromatid Exchanges*. Part A. The Nature of SCEs. Eds. R.R. Tice & A. Hollaender, Plenum Press, pp. 481–91.

Scott, D. (1982). UKEMS collaborative genotoxicity trial. Cytogenetic tests of chloromethylbiphenyl, BC and 4HMB. Summary and appraisal. *Mutation Research*, **100**, 313–31.

Snedecor, G.W. & Cochran, W.G. (1980). *Statistical Methods*. 7th Edn. Iowa State University Press, Iowa.

Wulf, H.C., Husum, B., Engberg-Pederson, H. & Niebuhr, E. (1984). Guidelines for the statistical evaluation of SCE. In *Sister Chromatid Exchanges*. Part A. The Nature of SCEs. Eds. R.R. Tice & A. Hollaender, Plenum Press, pp. 441–55.

7

Statistical analysis of *in vivo* cytogenetic assays

D.P. LOVELL D. ANDERSON (Group Leader)
R. ALBANESE G.E. AMPHLETT
G. CLARE R. FERGUSON
M. RICHOLD D.G. PAPWORTH
J.R.K. SAVAGE

7.1 INTRODUCTION

The Department of Health and Social Security (DHSS) through its Committee on Medical Aspects (COMA) has provided guidelines on the testing of chemicals for mutagenic potential. One of the tests is a demonstration of 'induced chromosome damage in the intact animal either using the micronucleus test or, preferably, metaphase analysis of bone marrow or other proliferative cells' (DHSS, 1981). Topham *et al.* (1983) outlined the test procedures for the performance of these tests. Full details of the suggested protocols, preparation of the test material, selection of animals, their husbandry and other such factors are discussed there.

This chapter extends the discussion to the statistical analysis of data obtained from such tests and its interpretation. The analysis of micronucleus test data will be described first as the problems involved are somewhat simpler. The issues arising from the analysis of cytogenetic data will be discussed later. Further discussion of Mammalian Germ Cell Cytogenetics can be found in Albanese *et al.* (1984).

7.2 THE MICRONUCLEUS TEST

The micronucleus test has become a widely used screening procedure for the detection of potential chromosome damaging agents and/or spindle poisons (Heddle *et al.*, 1983). Micronuclei are usually analysed in the *newly formed* polychromatic erythrocytes (PE) in rodent bone marrow; they are isolated or broken chromosome fragments which are not expelled when the nucleus is lost from the normoblast during its last

Table 7.1. *Guidelines for* in vivo *cytogenetic assays adapted from UKEMS Guidelines Part 1*

Micronucleus test	Rodent bone marrow metaphase assay
1. At least five male and five female animals per experiment and control group.	1. At least five male and five female animals per experimental and control group.
2. A single dose of the test compound at either the maximum tolerated dose (MTD) or dose producing some cytotoxicity. Cells should be sampled at three time points between 12 and 72 hours after treatment.	2. A single administration should be used in preference to multiple dosing.
3. At least 1000 polychromatic erythrocytes (PEs) should be examined for micronuclei in each animal.	3. A basic screen can consist of one dose of a compound. This dose should be the maximum tolerated dose (MTD) or one producing some evidence of cytotoxicity. Animals should be sampled at three time points between 6 and 48 hours after administration and the second sampling should be at 24 hours.
4. Additional doses used, for instance, in verification studies, should be sampled at the most sensitive interval or if not known, at 24 hours.	4. A verification study requires additional doses; sampling can be at a single interval (the most sensitive, or if this is not known at 24 hours).
	5. At least 50 well-spread complete metaphase cells per animal should be scored for chromosomal aberrations.

cell division but remain in the cytoplasms forming micronuclei (MPE). The presence of such micronuclei provides a measurement of chromosome damage or spindle malfunction in somatic cells indicating that a compound may have clastogenic effects.

A number of statistical issues are raised in the selection of appropriate dose levels, numbers of dose levels, animals per dose and cells to be scored per animal. The sampling time selected and whether single or multiple dosing regimes are used also affect the analysis. The way the data is collected also has statistical implications. Ashby & Mohammed (1986) have suggested, for instance, that different methods of slide preparation and assessment by different observers may invalidate some studies. They concluded that measures to reduce this source of variability will need to be incorporated in future protocols. They also suggest the number of cells scored should be increased from 1000 to 2000/slide.

The micronucleus test procedures recommended by the United Kingdom Health & Safety Executive (HSE), the European Economic Community (EEC) and the Organisation for Economic Co-operation & Development (OECD) share the common properties listed in Table 7.1.

Topham *et al.* (1983) suggested that at least five animals should be used for each sampling point with a minimum of four male and four female if both sexes are used. The background frequency of micronucleated cells suggested by these authors should not be more than 0.4% (4/1000). The proportion of polychromatic to normochromatic cells should be between ·0.4 and 1.0.

Topham *et al.* (1983) stated that studies conforming to the design outlined above have the power to detect the doubling of the background incidence of 0.4% with 95% confidence. (This presumably means that the test would have a power of 95% at the α level of 5% although no details were given of the calculations carried out to determine these effects.) The introductory chapter explains the theory behind significance levels and power. Calculations based upon the methods developed by Margolin *et al.* (1983) confirm the power calculation. Evaluation of the data should include statistical tests and a positive response was defined as one where at least two treatment groups are significantly positive at $P<0.05$. It was stressed that such increases in micronuclei 'should bear a biologically sensible relationship to each other' (Topham *et al.*, 1983, p.136).

The choice of group sizes of about 10 in the various guidelines are based mainly upon experience, accepted practice, convenience and a feeling that 'the numbers are about right'. Work in progress is likely to show that such designs do have a high chance of detecting effects resulting in a doubling of the control incidence; that 2000 PEs should ideally be scored per animal and that, in the absence of appreciable between-animal variability, increasing the number of PEs scored could be matched with a reduction in the numbers of animals per group.

7.2.1 The raw data

The basic measurement for analysis is the proportion of micronucleated cells (MPEs) in a series of PEs. *The experimental unit is the animal not the cell.* The data are expressed either as numbers of micronucleated cells or as percentages and difficulties in the statistical analysis arise because these counts are small values (often zero or one). There is the further problem of possible heterogeneity (variability) between animals in the same dose groups.

7.2.2 Approaches to the analysis of the data

The type of data collected is not particularly unusual despite the points made in the previous section. Various methods of analysis have been used by researchers employing the micronucleus test. Parametric statistics, i.e. statistics usually based upon the normal distribution, have been used after transformation of the data; non-parametric methods based upon relative rankings have been applied and methods based upon contingency tables have also been used.

An unpublished report by the Statisticians in the Pharmaceutical Industry (PSI, 1986) examined the performance of four different approaches to the analysis of a series of micronucleus studies conforming to OECD guidelines. These methods – a non-parametric, a parametric after transformation, a likelihood ratio test based upon the Poisson distribution and a dose-response study using general linear modelling methods – all gave similar results.

Three sets of data (Tables 7.2–7.4) conforming to various aspects of the OECD guidelines were used for the analyses demonstrated below, and were chosen to illustrate specific points and likely difficulties in interpretation.

7.2.3 The distribution of micronucleus scores

Micronuclei are either present or absent in a PE. They can therefore be expressed as either the number or the proportion of PE with micronuclei. Their relative rarity in control animals suggests that their incidence follows a Poisson distribution, i.e. rare independent events occurring in large sample sizes. Studies on the incidence of micronuclei in control animals have in general shown that a Poisson distribution gives a good approximation to the data.

Data on the incidence of micronuclei in control mice from five years of investigation in one laboratory are provided in Box 7.1. Two different methods of testing whether the data fits a Poisson distribution can be used. Firstly an analysis using the χ^2 test of goodness of fit (Snedecor & Cochran, 1980, p.131) suggests that the Poisson distribution was a reasonable fit. However, a second test, the dispersion test (Snedecor & Cochran, 1980, p.198), detected a significant excess of animals with higher micronuclei scores than expected based upon a Poisson distribution. The contradiction between these two tests is probably because the dispersion test is more sensitive at detecting 'over-dispersion' of data which is likely if the historical control data are the aggregation of a number of control groups each with a slightly different underlying background incidence. In practical terms though, the data are sufficiently close to that expected from a Poisson distribution to provide reassurance about their use as data on the background control incidence. (Exact tests of fit to a Poisson distribution are available (Papworth, 1983).)

Any of these tests could be used to check the distribution of the control incidence of micronuclei. The main consideration is an examination of the background control values, showing them to be within acceptable limits and not fluctuating greatly over time, rather than formal hypothesis testing.

Mackey & MacGregor (1979) have suggested that another statistical

Table 7.2. *Data set A*.

		Sampling time					
		24 h		48 h		72 h	
		m	*p/n*	*m*	*p/n*	*m*	*p/n*
Vehicle (negative) control	Male	1	0.808	0	1.061	0	1.275
		0	0.772	1	1.187	0	1.140
		1	0.985	1	0.963	0	1.030
		0	0.925	0	1.065	0	1.127
		1	0.677	0	1.145	0	0.994
	Female	1	0.864	0	0.807	0	0.785
		1	0.786	0	0.934	1	1.330
		0	0.715	0	0.862	3	1.077
		1	0.810	0	0.910	1	0.758
		1	0.987	1	0.918	1	1.319
Test substance 'V'	Male	1	0.457	1	0.599	Died	—
		1	0.793	Died	—	0	0.927
		0	0.653	Died	—	Died	—
		0	0.811	1	0.717	0	0.750
		3	0.168	1	0.027	1	0.329
	Female	0	0.584	0	0.507	1	0.562
		0	0.802	0	0.119	0	0.238
		1	0.742	Died	—	1	0.190
		0	0.423	1	0.309	Died	—
		0	0.485	0	0.076	1	0.297
Positive control (mitomycin C)	Male	94	0.680				
		38	0.056				
		78	0.736				
		1	0.766				
		84	0.334				
	Female	112	0.632				
		39	0.067				
		38	0.503				
		39	0.317				
		61	0.744				

Notes:
m = number of micronucleated polychromatic erythocytes out of 1000 polychromatic erythrocytes examined.
p/n = ratio of polychromatic to normochromatic erythrocytes.

Table 7.3. *Data set B.*

	m	*p/n*
Vehicle (negative) control		
Male	1	0.330
	0	0.979
	1	0.795
	6	0.994
	1	0.855
Female	0	0.933
	2	0.700
	1	1.274
	1	0.974
	2	1.011
'W' 10 mg/kg		
Male	0	0.442
	0	0.425
	0	0.913
	1	0.719
	2	0.881
Female	2	0.944
	0	1.240
	0	1.130
	1	0.598
	2	0.558
'W' 50 mg/kg		
Male	0	0.676
	1	0.860
	1	0.396
	0	0.929
	2	1.039
Female	1	1.024
	1	0.613
	0	0.550
	0	0.703
	3	0.564
'W' 250 mg/kg		
Male	1	0.747
	5	0.867
	1	0.357
	0	0.984
	1	0.885
Female	0	0.537
	0	0.613
	1	1.186
	0	1.216
	3	0.737
'W' 1250 mg/kg		
Male	2	0.494
	3	0.958
	4	1.325
	5	0.188
	14	1.293
Female	2	0.690
	0	0.752
	1	0.996
	2	0.966
	2	0.418
'W' 6250 mg/kg		
Male	16	0.345
	39	0.069
	25	0.279
	62	0.175
	37	0.573
Female	68	0.221
	34	0.494
	37	0.171
	45	0.205
	14	0.482
Reference substance (positive control)		
Male	3	1.028
	4	0.642
	10	0.606
	3	0.581
	12	0.359
Female	10	0.449
	14	0.544
	2	1.062
	11	0.371
	14	0.705

Notes:
m = number of micronucleated polychromatic erythrocytes out of 1000 polychromatic erythrocytes examined.
p/n = ratio of polychromatic to normochromatic erythrocytes.

Table 7.4. *Data set C.*

	m	*p/n*
Vehicle (negative) control		
Male	1	1.190
	3	0.582
	2	1.056
	2	1.075
	1	0.897
Female	2	1.064
	4	1.599
	2	1.126
	2	1.468
	2	1.480
'X' 67 mg/kg		
Male	5	0.381
	4	0.900
	1	1.071
	0	0.387
	2	1.022
Female	0	1.111
	4	0.945
	5	1.128
	1	0.917
	3	1.506
'X' 201 mg/kg		
Male	4	0.604
	5	1.254
	8	1.258
	5	0.757
	8	0.825
Female	2	0.544
	1	0.811
	4	0.495
	3	0.990
	8	1.105

Notes:
m = number of micronucleated polychromatic erythrocytes out of 1000 polychromatic erythrocytes examined.
p/n = ratio of polychromatic to normochromatic erythroyctes.

Box 7.1. *Historical control micronucleus data collected over 5 years of use of the assay in a single laboratory with a single stock of mice*

Incidence of micronuclei/ 1000 cells	0	1	2	3	4	5	6	7	>7	Total	Mean ($\overline{X}$)	Variance (SD^2)
Number of animals observed	944	853	400	110	31	8	3	2	0	2351	0.9268	0.9895
						13 (5–7 combined)						
expected	930.6	862.5	399.7	123.5	28.6	5.3	0.8	0.1	0			
						6.2 (5–7 combined)						

Test of Goodness of fit with Poisson distribution: $\chi_4^2 = 9.28$ $P = 0.10$–0.05
The observed and expected frequencies for five or more micronuclei have been combined in accordance with Snedecor & Cochran's recommendation (Method – Snedecor & Cochran, 1980, p. 131–2)
Test of Poisson distribution using dispersion test: $\chi_{2350}^2 = 2350\ SD^2/\overline{X} = 2508.98$
(Method – Snedecor & Cochran 1980, p. 198)
To relate this to probability tables, convert to normal deviate (z) by equation 9.3.1. (Snedecor & Cochran, 1968, p. 233).

$$z = \sqrt{(2\chi^2)} - \sqrt{(2d.f. - 1)} = \sqrt{(2 \times 2508.98)} - \sqrt{(2 \times 2350 - 1)} = 2.288$$

This has $P < 0.02$ from tables of normal frequency distribution (one-sided test) distribution, the negative binomial, may be a better fit to micronucleus data especially where there are increased incidences because of treatments. They have suggested an alternative experimental design based upon a sequential test which is claimed to be statistically more powerful and uses fewer animals. However, this design is considerably different from the standard OEÇD tests (although it would not necessarily be unacceptable to the regulatory authorities in the UK) and will not be considered further here.

7.2.4 The data sets

In this chapter, three sets of data (*A*, *B* and *C*) from a commercial laboratory have been used to illustrate different analyses that may be carried out. Each of the sets is used to illustrate one possible method of analysis of data from micronucleus studies. The data sets are shown in Tables 7.2–7.4.

7.2.5 A parametric approach using the analysis of variance

The first data set (*A*) consists of the micronucleus counts from five male and five female animals sampled 24, 48 and 72 hours after an acute dose of either a test substance '*V*' or the solvent used as the vehicle (the vehicle control). A positive control of five male and five female animals was also sampled 24 hours after injection of mitomycin C.

The data consists of the number of micronucleated cells in a sample of 1000 PEs and the ratio of polychromatic to normochromatic erythrocytes for each animal. A total of six mice (four male and two female) in the test group died during the course of the experiment and thus no data were available from them.

The factorial analysis of variance is a statistical technique which allows the effects of a number of different factors, (in this case, sex, treatment, time of sample), to be examined simultaneously. The underlying assumption behind the analysis of variance, and other parametric methods like the *t*-test or linear regression, is that the subsets of data within each combination of factors are normally distributed with equal variances. The counts of MPEs or their proportions are discontinuous data and do not meet these assumptions. Also their variances are not equal as they are related to the mean values. As a result the distributions are often skewed. A number of transformations are possible to overcome these violations of the assumptions and to allow the counts and proportions to be analysed by parametric methods. Tests of normality and constancy of variance are possible though in practice difficult where each combination consists of small numbers. Such tests are not usually necessary but if in doubt about the adequacy of the data consult a professional statistician. Parametric tests such as the analysis of variance are however quite robust so that even if all the assumptions are not totally met the results produced are not grossly affected.

In this analysis, problems arise because many of the control data are either zero or very low scores. There are also missing values because some of the treated animals died (if there are many missing values in the data the analysis should be carried out in conjunction with a professional statistician). The data were therefore analysed by a statistical package called Genstat* (Alvey *et al.*, 1982) using a square root transformation often used to 'normalise' counts. In fact, the transformation $\sqrt{(x+1)}$ is often used to stabilise the variances more effectively.

Box 7.2 shows the instructions used to provide the analysis using Genstat. (Genstat is available on many larger computer systems but other statistical packages could be used to produce similar analyses. A professional statistician may need to be consulted to help run the packages.)

Box 7.3 shows the analysis of variance table produced by the Genstat analysis. Values are estimated for the six animals in the treated groups who died. These in fact are the mean values for the survivors in the relevant groups. An analysis of variance table is produced which reflects the experimental design used and partitions the total variation into that resulting from the factors and the interactions between them. The variation resulting from the factors is compared with the variation between animals within the combinations (residual error) by obtaining the ratio of their respective mean squares (*MS*) with the residual *MS* (the variance ratio, *VR*). The resulting *VR* values may be compared against known values of *VR* (also known as *F*) either from tables or from values calculated by the package. The larger the *VR* value the less likely that variation between means is due to chance alone.

There was no difference in micronucleus counts between the control and the test groups, between the sexes or between the sample times. Only two sources of variation gave significant values: the interaction between sex and treatment ($P<0.05$) and that between sex and sample time ($P<0.01$).

Examination of the table of means (Box 7.4) shows that the first significant interaction is a consequence of the females having higher micronuclei scores in the control group but lower counts in the treated group than the males. The statistical significance probably arises from Genstat estimating missing values by the mean of the remaining animals in a group. This exaggerates the differences between the sexes at 48 hours where all the surviving males had scores of 1.

* See Appendix 2 for a discussion of statistical packages of use in analysing mutagenicity data.

Box 7.2. *Genstat instructions (lines 1–18) to run factorial analysis of variance on Data set A.*

```
GENSTAT V      RELEASE 4.04A
COPYRIGHT 1982 LAWES AGRICULTURAL TRUST (ROTHAMSTED EXPERIMENTAL STATION)

 1 'REFE'  MICRONUCLEUS_1
 2 'OUTPUT'  1,80
 3 'UNIT'  MOUSE $60
 4 'NAME'  SN=MALE, FEMALE
 5 :          DN=CONTROL, TREATMENT
 6 'VARIATE'  HN=24,48,72
 7 'FACTOR'  SEX $SN=5(1,2)6                         Definition of factors, specification of
 8 :           HOUR $HN=20(1 . . .3)                 names and order of data input
 9 :           DOSE $DN=10(1,2)3
10 'INPUT'  2                                        Read ratio of polychromatic to normochromatic, and
11 'READ' PROP,MNO                                   micronucleus number for each mouse from separate
12 'INPUT'  1                                        data file
13 'CALC'  TMNO=SQRT(MNO+1)                          Various square root-based transformations of
14 :          T1MNO=SQRT(MNO+0.5)                    micronucleus count
15 :          T2MNO= -SQRT(MNO)+SQRT(MNO+1)
16 'TREATMENT'  SEX*DOSE*HOUR                        Specification of factorial design
17 'ANOVA/PROB=Y'  PROP,TMNO,T1MNO,T2MNO,MNO         Call for analysis of variance to be carried out on
18 'RUN'                                             various tranformations of data

IDENTIFIER  MINIMUM  MEAN    MAXIMUM  VALUES  MISSING        { Data description. Note distribution
PROP        0.0270   0.7515  1.3300   60      6              { of micronucleus data is indicated
MNO         0.0000   0.5556  3.0000   60      6        SKEW  { to be skewed
```

Box 7.3. *Anova table on transformed micronucleus counts*

***** Analysis of variance *****

Variate: TMNO $\sqrt{(X+1)}$	*Degrees of freedom (missing values)*	*Sum of squares*		*Mean squares*	*Variance ratio*	*Probability of variance ratio*
Source of variation	D.F (MV)	SS	SS%[a]	MS	VR	F PR
*Units *Stratum						
Sex	1	0.000 33	0.01	0.000 33	0.006	0.939
Dose	1	0.007 86	0.23	0.007 86	0.139	0.711
Hour ≈ *Time of sampling*	2	0.043 34	1.25	0.021 67	0.383	0.684
Sex.dose	1	0.314 47	9.04	0.314 47	5.553	0.023*
Sex.hour	2	0.723 46	20.81	0.361 73	6.387	0.004**
Dose.hour	2	0.103 51	2.98	0.051 75	0.914	0.409
Sex.dose.hour	2	0.012 41	0.36	0.006 21	0.110	0.896
Residual	42(6)	2.378 64	68.41	0.056 63		
Total	53	3.584 03	103.07	0.067 62		
Grand total	53	3.584 03	103.07			

Estimated grand mean	1.224
Total number of observations	60
Number of missing values	6
Maximum number of iterations	3

[a] *% of sum of squares accounted for by particular item sex, dose etc. Adds up to > 100% because of bias introduced by missing values*

$^{*}P<0.05$
$^{**}P<0.01$

Unit number	Estimated value
32	1.411
33	1.411
38	1.104
51	1.139
53	1.139
59	1.311

Values estimated from data for six animals that died in test

Box 7.4. *Means of transformed micronuclei counts from Data set A produced by Genstat*

***** Tables of means *****

Variate: TMNO

Grand mean	1.224		

Sex	Male	Female
	1.222	1.227

Dose	Control	Treatment
	1.213	1.236

Hour	24.00	48.00	72.00
	1.257	1.191	1.224

Dose Sex	Control	Treatment
Male	1.138	1.306
Female	1.288	1.166

Hour Sex	24.00	48.00	72.00
Male	1.307	1.289	1.069
Female	1.207	1.093	1.380

Hour Dose	24.00	48.00	72.00
Control	1.290	1.124	1.224
Treatment	1.224	1.258	1.225

Dose	Control			Treatment		
Hour Sex	24.00	48.00	72.00	24.00	48.00	72.00
Male	1.249	1.166	1.000	1.366	1.413	1.139
Female	1.331	1.083	1.449	1.083	1.104	1.311

All combinations of means are shown together with an estimate of the standard errors of the differences between any two means in a particular table.

Note that the standard errors are approximate because they are not adjusted for missing values.

***** Standard errors of differences of means *****

Table	Sex	Dose	Hour	Sex dose	Sex hour	Dose hour	Sex dose hour
REP	30	30	20	15	10	10	5
SED	0.0614	0.0614	0.0753	0.0869	0.1064	0.1064	0.1505

The other significant interaction, that between sex and sampling time, results from females having lower scores than males at 24 and 48 hours but higher scores at 72 hours. Of the males in the 72-hour group seven out of eight had scores of 0 while seven out of nine females had scores of 1 or more. This may represent a chance effect although the probability of such a finding is, based upon a two-sided Fisher's exact test, $P = 0.024$ (calculated by doubling the one-sided P value). A significant interaction of this nature suggests a check should be carried out that the experimental design and data collection were carried out according to protocol. Possible causes, besides chance, are male and females tested at different times or scored by different people.

Overall the conclusion is that the compound did not affect the incidence of micronuclei in the bone marrow *PE*s of treated mice compared with control mice. No statistical tests need to be carried out to show that the positive controls differed from both the vehicle control and the treated group.

A similar analysis was carried out on the proportion of polychromatic to normochromatic cells which provides an estimate of the general toxicity of the chemical to the bone marrow.

There was a highly significant treatment effect ($P < 0.001$) and a significant interaction between treatment and sample time ($P < 0.01$). The proportion of polychromatic to normochromatic cells was reduced at all times of sampling, particularly at 48 hours, showing that the chemical was producing toxic effects in the bone marrow. (There was also a significant sex difference with males having slightly higher ratios: 0.79 versus 0.66 $P < 0.05$.)

It can be seen, therefore, that the factorial analysis of variance allows a comprehensive analysis of a complex data set. It provides powerful statistical tests of the effect of the treatment as well as tests of interactions between the factors included in the design. In this example the use of a square root transformation produced a satisfactory result even though the residual mean square value was considerably lower than the theoretical mean square value of 0.25 expected after such a transformation. This is a consequence of the low scores which included a number of zeros and will affect to some extent the Type I and Type II error rates (see Chapter 1).

The data were also analysed by a series of other tranformations such as $\sqrt{X}, \sqrt{(X+0.5)}, \sqrt{(X+1)}+\sqrt{X}$. All gave similar results as did analysis with the untransformed data (which is not recommended). A further possible transformation is the inverse hyperbolic sine which has been used by Mitchell & Brice (1986) and is suitable for data which is distributed as a negative binomial.

7.2.6 An approach using likelihood ratio tests

Amphlett & Delow (1984) have proposed that data from micronucleus studies can be analysed by tests based upon the Poisson distribution. The tests they use are based upon likelihood ratio tests and are applicable providing the data at each dose level and in the controls fit a Poisson distribution.

A test can be made of whether the data fit a Poisson distribution by using a modification of the dispersion test illustrated in Box 7.1. If the data fit a Poisson distribution then a likelihood ratio test can be carried out between the control and each of the test groups in turn.

We have used data set B to demonstrate the application of the likelihood-ratio test. Micronucleus counts at 24 hours were obtained from five male and five female animals at five doses of compound 'W' together with positive and negative controls. Micronucleus scores for the negative control and the first four dose levels are used. Data from the top dose (6250 mg/kg) and the reference control substance are not included because the effects are clearly positive.

In this case the dispersion test (Box 7.5) gave a highly significant ($P < 0.001$) result indicating that the data are over-dispersed compared with a Poisson distribution. Amphlett & Delow state that the dispersion test sometimes fails for potent carcinogens which then invalidates the likelihood ratio test. Here the compound is clearly positive based upon the results at 6250 mg/kg but we have continued with the likelihood ratio tests between the negative control and the remaining test groups.

All the comparisons are non-significant except at the 1250 mg/kg dose level (Box 7.5). The significance of the comparison is difficult to assess as a consequence of the significant dispersion test which means that the likelihood ratio test may overestimate the statistical significance (i.e. cause too many false positive or Type I errors; see Chapter 1). Examination of the 'raw' data in Table 7.3 suggests there is considerable heterogeneity between animals as well as a possible sex difference at 1250 mg/kg with males appearing more sensitive.

Amphlett & Delow also discuss the use of multiple comparisons methods so that the chance of a false positive (α or Type I error) is either 5% on an individual comparison basis or on an experiment-wide basis. A discussion of multiple comparison methods will be found in the Introductory Chapter.

7.2.7 An approach using generalised linear models

The likelihood ratio test developed by Amphlett & Delow is a special case of a more general method based upon fitting various statistical

Box 7.5. *Likelihood ratio test of micronuclei data from Data set B.*

1. *Dispersion test*

The equation is $\chi_t^2 = \sum_{i=1}^{k} \left[\sum_{j=1}^{n_i} \frac{(X_{ij} - \overline{X}_{i.})^2}{\overline{X}_{i.}} \right]$

where $t = \sum_{i=1}^{k} (n_i - 1)$, $X_{i.} = \sum_{j=1}^{n_i} X_{ij}$ and $\overline{X}_{i.} = X_{i.}/n_i$

For the data set B

Dose mg/kg	0	10	50	250	1250
n_i	10	10	10	10	10
$X_{i.}$	15	8	9	12	35
$\overline{X}_{i.}$	1.5	0.8	0.9	1.2	3.5
$\sum_{j=1}^{n_i} (X_{ij} - \overline{X}_{i.})^2$	26.5	7.6	8.9	23.6	140.5

$\chi^2_{(45)} = 17.667 + 9.500 + 9.889 + 19.667 + 40.143$
$= 96.866 \ (P < 0.001)$

2. *Comparison of test dose* v *negative control*

$\chi^2_{(1)} = 2[X_{1.} \ln \overline{X}_{1.} + X_{2.} \ln \overline{X}_{2.} - X_{..} \ln \overline{X}_{..}]$ where $X_{..} = \sum_{i=1}^{k} \sum_{j=1}^{2} X_{ij}$,
$\overline{X}_{..} = X_{..}/20$ and ln is the natural logarithm

for $X_{1.} = 15$, $X_{2.} = 8$ $X_{..} = 23 (\equiv 15 + 8)$
$= 2[15.\ln(1.5) + 8.\ln(0.8) - 23.\ln(1.15)]$
then $\chi^2_{(1)} = 2.16$ and $\sqrt{(\chi^2_{(1)})} = 1.47$
$\sqrt{(\chi^2)}$ can be used as a one-sided test using a table of the normal distribution
Similarly for
0 mg/kg *v* 50 mg/kg $\chi^2_{(1)} = 1.52$ $\sqrt{(\chi^2_{(1)})} = 1.23$
0 mg/kg *v* 250 mg/kg $\chi^2_{(1)} = 0.33$ $\sqrt{(\chi^2_{(1)})} = 0.58$
0 mg/kg *v* 1250 mg/kg $\chi^2_{(1)} = 8.23$ $\sqrt{(\chi^2_{(1)})} = 2.87$, $P < 0.01$
(These tests are described in more detail by Amphlett & Delow, 1984.)

models to sets of data. This method, called the Generalised Linear Model, allows factors such as sex, dose and sample times to be incorporated into the analysis. The resemblance to the approach using the analysis of variance is more than just superficial as the analysis of variance is a special case of the general linear model. Generalised Linear Models allow parametric approaches based upon models other than the normal to be used (such as the binomial) so avoiding the problem caused by a choice of a transformation.

The effects of factors are tested using a modification of the likelihood ratio test. A full description of the procedures and interpretation of such an analysis is beyond the scope of this chapter but the analysis can be programmed using the statistical package GLIM (Generalised Linear Interactive Modelling; Baker & Nelder, 1978) or by using procedures in the Genstat package. Analyses using generalised linear models are unlikely to be done without the assistance (and probably supervision) of a professional statistician but are likely to be satisfactory methods for the analysis of micronucleus data.

7.2.8 An approach using 2 × 2 contingency tables

Some statisticians feel uneasy about the use of a transformation such as the square root transformation used in data set A when there are a large number of low scores. An alternative approach based upon contingency χ^2 tables is possible.

Data set C consists of the micronucleus scores from five males and five females treated with the vehicle control, 67 or 201 mg/kg of 'X' with sampling carried out 30 hours after dosing.

The first stage in this analysis is to test for evidence of heterogeneity between animals within the dose groups (i.e. is there appreciable between-animal variation?). This is done by calculating the heterogeneity χ^2 for the differences between the proportions for each animal (Box 7.6).

In this example all the groups were homogeneous and comparisons can be made between the control and test groups by pooling over individual animals. The test is made by using Pearson's contingency 2 × 2 χ^2 test with 1 degree of freedom. This usually includes a correction factor named after Frank Yates who first suggested it in 1934 (Yates, 1934). This gives a more accurate estimate of the χ^2 value especially when some of the numbers in the 2 × 2 table are small. Box 7.7 shows the analysis for two of the comparisons.

There was no significant difference between the 0 and 67 mg/kg dosed animals but a highly significant increase in frequency between 0 and 201 mg/kg ($P < 0.01$).

An alternative way of testing for significance levels in a 2 × 2 table is by calculating exact probabilities using Fisher's exact test. Large sample sizes however can result in considerable computation and this test requires a computer. Exact tests are also possible for tests of heterogeneity in $m \times 2$ contingency tables (Box 7.6). Again these require access to specific programs and a powerful computer.

An alternative test to pairwise comparison as carried out by 2 × 2 tables

Box 7.6. *Heterogeneity chi-square (Data set C).*
(This is similar to tests of heterogeneity of variances in parametric analyses.)

The equation is

$$\chi^2_{(m-1)} = \frac{n^2}{r(n-r)} \left[\sum_{i=1}^{m} \frac{r_i^2}{n_i} - \frac{r^2}{n} \right]$$

where m is the number of animals in the group,
r_i and n_i are the number of micronucleated cells and the number of cells for the ith animal respectively,
and
r and n are the total number of micronucleated cells and total cells over all animals.
Substituting the values for the vehicle control group this gives

$$\chi^2_{(9)} = \frac{10000^2}{21 \times 9979} \left[\frac{1^2}{1000} + \frac{3^2}{1000} + \ldots + \frac{2^2}{1000} - \frac{21^2}{10\,000} \right]$$

$\chi^2_{(9)} = 3.29$ $P > 0.95$ (from a table of the χ^2 distribution)
For the 67 mg/kg group
$\chi^2_{(9)} = 13.83$ $P = 0.25 - 0.10$
and for the 201 mg/kg group
$\chi^2_{(9)} = 12.06$ $P = 0.25 - 0.10$
In this case all three sub-groups are homogeneous and comparisons can be made between the vehicle control and the dose groups using 2×2 contingency tests.

in Box 7.7 is to test for a linear trend. This procedure, which is comparable to linear regression in parametric studies, tests for increasing frequency with increasing dose. It requires giving values to the different doses either by using the original scale, or converting to log dose (problems arise when converting 0 to logs!) or ranking the doses 0, 1, 2. Box 7.8 shows the calculations; in this case the original dose values are used. There is evidence of an increasing proportion of micronuclei with increasing dose. The findings from the 2×2 tables is that this is particularly so between the groups dosed at 67 and 201 mg/kg although there was no evidence for deviation from the regression line.

In general, the control groups are likely to be homogeneous and it is only in the test groups where an effect has occurred that heterogeneity may be expected. If heterogeneity is found it is possible to carry out statistical tests which have a similar underlying structure and philosophy to the modified *t*-tests when there is significant heterogeneity of the within-group variances. It is also possible to pool over a series of sample times by a test called Cochran's test. Full details of this test can be found in Snedecor & Cochran (1980, p. 212) where methods for the combination of sets of 2×2 table are described.

Box 7.7. *The 2 × 2 contingency table used on Data set C*
The data are set out as follows:

	Cells with micronuclei (M)	Cells without micronuclei (NM)	
Control cells (C)	a	b	$a+b$
Treated cells (T)	c	d	$c+d$
	$a+c$	$b+d$	

where $a+b+c+d=N$

A common formula for calculating χ^2_c (which has one degree of freedom) is $\chi^2_c = \dfrac{N(|ad-bc|-N/2)^2}{(a+b)(c+d)(a+c)(b+d)}$ Snedecor & Cochran (1968) equation 8.10.3, p. 217)

This includes the continuity correction as the data are discrete. The equation is widely published in many slightly different forms. (The symbols || indicate the absolute value of $ad-bc$, i.e. irrespective of sign.)

Substituting the values of control v 67 mg/kg 'X'

	M	NM
C	21	9979
T	25	9975

$$\chi^2_c = \frac{20000(|-40\ 000|-10000)^2}{10000 \times 10000 \times 46 \times 19954}$$

$= 0.20$ $P = 0.70–0.60$ (from tables of χ^2 distribution)

Similarly for control v 201 mg/kg 'X'

	M	NM
C	21	9979
T	48	9952

$\chi^2_c = 9.83$ $P < 0.01$

7.2.9 Discussion of the various approaches to analysis of micronucleus data

The three approaches demonstrated here, in general, all produced similar conclusions when they were applied in turn to the three data sets A, B and C. This is in broad agreement with the conclusions reached in the report of Statisticians in the Pharmaceutical Industry. Other methods have been proposed for the analysis of the data; in particular, non-parametric

Box 7.8. *A test for a linear trend in proportions used on Data set C.*

Dose (D_i)	Proportion of PCEs with micronuclei $p_i = r_i/n_i$
$D_1 = 0$	$p_1 = r_1/n_1 = 21/10000$
$D_2 = 67$	$p_2 = r_2/n_2 = 25/10000$
$D_3 = 201$	$p_3 = r_3/n_3 = 48/10000$
Total	$p = r/n = 94/30000$

A regression coefficient b is calculated by the following equation (see Snedecor & Cochran, 1980, p. 206–7).

Numerator $= \sum r_i D_i - (\sum r_i)(\sum n_i D_i)/\sum n_i$
$= 2925.67$

Denominator $= \sum n_i D_i^2 - (\sum n_i D_i)^2/\sum n_i$
$= 209486666.7$

this gives $b = 1.39659 \times 10^{-5}$

Its standard error $s_b = \sqrt{(pq/\text{denominator})}$ where $q = (1-p)$.
$= 3.861398 \times 10^{-6}$

This is tested against the normal deviate
$z = b/s_b = 3.617$ $P < 0.002$ (from a table of the normal distribution)

This result can also be expressed in an analysis of variance style table where the overall chi-square between doses is split into a linear and deviation from linear trend component. The table for the data in this example is reproduced below. Details of how to calculate the respective chi-squares are given in Everitt, 1977, p. 51–6.

Source	*d.f.*	Chi-square	*P*
Linear regression	1	13.08	<0.001
Deviation	1	0.52	0.50–0.40
Overall chi square	2	13.60	

methods. The analysis using the non-parametric equivalents of the analysis of variance and two sample *t*-test, the Kruskal–Wallis and Mann–Whitney *U*-tests, appear feasible. The non-parametric equivalent of the linear regression test for trend, Jonckheere's test, could also be used.

It has been suggested that the distribution of the micronucleus counts as a negative binomial rather than a Poisson, especially in the test groups, invalidates the assumptions behind many statistical tests (Mitchell *et al.*, 1981). Thus, micronucleus data should be analysed using another non-parametric test, i.e. the two-sample Kolmogorov–Smirnov test (Mitchell & Brice, 1986). This compares both the average value of the samples and their distribution. In Mitchell & Brice's study of its performance on sets of data it is the most conservative (i.e. gives the lowest number of significant results) of a range of statistical tests. However, at the moment its use is not recommended because little is known of its power (ability to detect real effects) in studies of micronucleus data. Campbell (1974) suggests that the scale of values obtained should be divided into divisions with three or less observations per division and that coarser subdivision reduces the power of the test. In micronucleus studies where values of 0 and 1 are common, adequate subdivision of the data may be difficult, resulting in a bias in the conclusions that are drawn.

The choice of a statistical method to analyse micronucleus data will depend upon personal preference as well as access to computing and statistical facilities and expertise. Methods based upon the χ^2, likelihood ratio or non-parametric tests may be easier to do if such facilities are limited; when they are available, analysis of variance or generalised linear modelling approaches may be preferred.

In general, methods which have high statistical power such as tests for dose-related trends, are preferable especially when they help to identify areas where biological expertise is required to help with the interpretation. Methods taking into account the multiple possible comparisons between groups may be applied to reduce the false positive rate. A number of different methods exist (see, for instance, Snedecor & Cochran, 1980, p. 233). Their use should be clearly stated since the consequences are an increase in the false negative rate (i.e. a loss of statistical power) of a test.

Significance tests of positive against negative (vehicle) controls are not strictly necessary although often demanded by referees and assessors for 'statistical sanctification'. If done a two-sample *t*-test, χ^2 or Mann–Whitney *U*-test are all probably acceptable. Positive control data should probably be excluded from more complex analyses such as the analysis of variance as the increased variation associated with a strong clastogenic agent may distort some of the analyses.

There is, of course, no end to the ingenuity of people to apply statistical methods inappropriately! Suggesting incorrect analyses to avoid is therefore difficult. Two approaches which should be avoided are (i) to analyse the test and control data by a *t*-test without transforming the data and (ii) to use the tables produced by Kastenbaum & Bowman (1970) without checking whether heterogeneity between animals is present.

7.3 *IN VIVO* CHROMOSOMAL ASSAYS

7.3.1 The rodent bone marrow metaphase assay

The metaphase analysis of bone marrow or other proliferative cells is the preferred test for the induction of chromosome damage in the intact animal in the DHSS's guidelines for the testing of chemicals for mutagenicity (DHSS, 1981).

In the most common version of the test rodents are dosed with the test compound, killed at predetermined times and metaphase spreads made of the bone marrow cells. These metaphase spreads are scored for various categories of chromosome and chromatid-type aberrations. The presence of small non-staining regions (achromatic lesions or 'gaps') is noted.

7.3.1.1 *Standard protocol*

A number of standard protocols are available. Those of the OECD, EEC and HSE have common criteria which are summarised in Table 7.1.

These guidelines are slightly different from those given by Topham *et al.* (1983) who suggest that at least 500 cells should be scored per dose with at least 50 (preferably 100) cells per animal. Topham *et al.* (1983) also suggested options for the choice of dose intervals in a dose–response study. These are (a) The Maximum Tolerated Dose (MTD), 0.5 MTD, 0.25 MTD, (b) MTD, 0.2 MTD, 0.1 MTD or (c) MTD, 0.2 MTD, 0.04 MTD.

Topham *et al.* (1983) discuss the importance of sample times. In particular, there is the possibility that clastogenic agents will be cytotoxic, reducing the number of cells entering mitosis as well as altering the cell cycles of those that do divide.

7.3.1.2 *The raw data*

Each metaphase spread is scored by an experienced observer for evidence of various types of structural aberrations and numerical changes. The types of damage and their classification are illustrated in Scott *et al.* (1983). Each cell can therefore be considered to have a score on a set of variates representing different types of damage. Such data are usually referred to as multivariate data. In general the spontaneous incidence of most of the aberrations is very low so that the values for most cells will be zero. Various suggestions have been made as to how to combine the various aberration types into aggregate combinations which are easier to analyse. Topham *et al.* (1983) recommended scoring cells for deletions and exchanges separately for both chromosome and chromatid types. The data sets analysed here were collected in accordance with these recommendations.

7.3.1.3 *Previous recommendations for analysis*

Topham *et al.* (1983) discussing *in vivo* cytogenetic assays, made no clear recommendations of statistical analyses to use except to suggest that they should be identified before the study started. Criteria for positive effects were outlined which included: a minimum increase over historical controls; the presence of a dose-related increase with at least two dose levels having significantly increased aberration frequencies compared with control values; the lesion seen must be consistent with damage expected relative to the time of dosing. Scott *et al.* (1983) recommended the use of either a χ^2 test or a Fisher's exact test between the control and each treated group in turn for *in vitro* data.

7.3.1.4 *Statistical considerations of the data – choice of measures*

Many of the considerations involved in the analysis of micronucleus data are applicable to *in vivo* chromosome aberration studies. There is potential variation between cells within an animal, between animals within a test group and as a result of the treatments. The multivariate nature of the data is an added complexity providing initially a more complex data set than the micronucleus data.

Cells may have several aberrations either of the same or different classes and a choice needs to be made whether to analyse by numbers of aberrations or by numbers of cells having one or more aberrations. The dose–response curves from these two methods will be different. Topham *et al.* (1983) felt that the proportion of cells expressing aberrations is more appropriate because of the difficulties of incorporating cells with multiple damage where this may be non-independent. (Margolin *et al.* (1986) came to a similar conclusion for *in vitro* data.)

As indicated earlier the multivariate nature of the data is simplified by grouping over various aberration types. The low spontaneous rate and the relatively small number of cells scored per animal would otherwise mean that the statistical tests would have low power for the detection of any treatment-related effects. The commonest measurement has therefore been to pool all aberrations and to analyse the proportion of cells showing aberrations.

There is considerable debate over the relevance of achromatic lesions or 'gaps' (Brøgger, 1982). These are a transitory phenomenon with no proven genetic significance and their relationship to structural chromosomal aberrations is unclear. It is usual to report the numbers observed so that an analysis can be performed with or without their inclusion. The interpretation of data which requires their inclusion to achieve statistical significance must be made with caution.

7.3.1.5 *Possible methods of analysis*

Statistical methods for the analysis of proportional data are usually based on the binomial distribution. The incidence of specific aberrations however may be related to an underlying Poisson distribution. The methods for the analysis of proportions are either the χ^2 tests, or the use of *t*-tests or analyses of variance after the data have been transformed by a suitable method such as the arcsin transformation (see Chapter 1).

In both types of analyses there is an underlying assumption that the variation between animals within each of the groups is homogeneous. (The tests also assume that the variation between cells within an animal conforms to either a binomial distribution or an underlying normal distribution after an arcsin transformation. In fact, chromosomal aberration data are frequently distributed as a negative binomial especially after exposure to chemical clastogens (Couzin & Papworth, 1979).)

Such tests are difficult to conduct when a large proportion of the values are zero or small. It is a matter of debate among statisticians as to how to proceed if significant heterogeneity is found within groups. A purist may argue that such heterogeneity invalidates further analysis; a pragmatist may reply that only gross distortions will severely affect the conclusions of tests like the analysis of variance. The problem is in obtaining a definition of 'gross'. When heterogeneity exists it will affect the incidence of false positive and false negative results. However in many cases, especially where the heterogeneity is found in test groups, this may not seriously distort the biological interpretation of the analysis. A reasonable approach seems to be to report any heterogeneity detected but to proceed with a normal analysis and allow the cytogeneticists their own choice of view on the subject.

Another potential difficulty is the problem of multiple comparisons. The number of possible comparisons depends upon the number of different types of aberration analysed. If a large number of possible statistical tests is carried out there is a high probability of obtaining false positives (statistical significance by chance alone). Various methods have been suggested for minimising the risks of multiple comparisons producing false positives. These are discussed in Chapter 1 so that these methods are not included in the analyses carried out here.

There is also the danger that in the aggregations of the raw data, certain low incidences of chromosomal aberrations (which nevertheless represent real biological effects) are submerged in the general 'noise'. The difficulty is that the background incidence is so low that even a small but non-significant increase in it would be biologically important. Chromosomal exchanges are, for instance, comparatively rare spontaneous events and their detection in treated material may be important. There is therefore a need for the 'raw' data and 'historical control values' to be presented as well as any modified data sets in the final report. Statistical tests do exist which incorporate information from background control data into formal statistical analyses. They require, though, some care in their interpretation as well as some confidence in the nature of the background control data.

7.3.1.6 *Examples of the use of the chi-square tests*

Two experimental data sets (D and E) will be used to illustrate aspects of the analysis of chromosomal aberration data. In each case χ^2 tests will be used, although approaches using the analysis of variance could be applied such as illustrated for the micronucleus assay. The raw data are presented in Tables 7.5 and 7.6.

The actual protocols do not agree totally with the recommended protocol described earlier but are sufficiently similar to illustrate both the analyses and the issues arising in their interpretation. The main difference is that the control and treated groups have 12 and 8 animals respectively. Such differences in sample sizes are unlikely to affect the effectiveness of the assays appreciably.

The first set of data (D) were chromosomal aberrations detected after various doses of compound 'Y'. Metaphase spreads were scored 6, 24 and 48 hours after treatment. The doses used were 40, 120 and 400 mg/kg. The protocol was similar to the standard except that eight animals were examined at each dose level and 12 in each control group. A positive control, ethyl methanesulphate (EMS), was administered to eight animals for each sample time. The data from the positive control were not included in the formal statistical analysis.

A test is first made of heterogeneity within each test and control group. The methodology is the same as that shown in Box 7.6. Table 7.7 shows the heterogeneity χ^2 for each dose group using the % cells with aberrations either including or excluding gaps. In a number of cases heterogeneity was found within the groups.

The heterogeneity χ^2 test is only an approximate test and becomes less satisfactory the smaller the expected frequencies. Originally, expected values of less than 5 were considered sufficient to seriously invalidate the test. It is now considered that the test is satisfactory in some cases even when expected values are as low as one. However, where the expected frequencies are clearly very low (as is the case with some of the individual types of aberration) the test is unsatisfactory and homogeneity has to be assumed to allow further analysis.

As stated earlier there are alternative ways of handling data after heterogeneity has been detected. For simplicity, heterogeneity will be noted here but standard 2×2 χ^2 contingency tests and linear trends tests as described in Box 7.7 and Box 7.8 are undertaken. Alternatively tests can be used which take into account the heterogeneity and make the analysis comparable to *t*-tests with heterogeneity of within group variances.

The χ^2 tests for a 2×2 design and a linear trend can be related to a normal distribution by taking the square root of the χ^2 value. This value, z or $\sqrt{(\chi^2)}$, has a positive value if the proportion of aberrations is higher in the treated group in the 2×2 test or the linear response is positive in the linear trend. [The equations given in Box 7.7 and Box 7.8 can be modified to obtain z directly without the need to calculate whether it is +ve or −ve.]

None of the comparisons showed a significant increase in any of the treated groups compared with the control at each of the sampling times. There was also no evidence of any significant positive linear trends (Box 7.9). The compound does not induce chromosome aberrations.

The second set of data (E) consists of two experiments of exposure to various doses of a compound 'Z' from 1 ppm to 1000 ppm. Metaphase spreads were made from bone marrow fixed 24 hours after exposure. In experiment 1 exposure was for 2 hours; in experiment 2 for 6 hours. Twelve rats were sampled from the control group and eight rats from each of the treated groups.

Some heterogeneity was detected within the groups particularly in the reanalysis of extra data in experiment 2. No attempt has been made to correct the χ^2 tests to take this heterogeneity into account. In experiment 1 there was evidence for a significant increase in the number of chromosome deletions at 100 ppm in tests against the controls (Table 7.8). No other individual tests were significant in a one-tailed 2×2 test. There was evidence for a highly significant linear trend with chromosomal deletions as well as all damage excluding gaps. Evidence of significant increases from control frequencies at two dose levels together with a highly significant linear trend would have met the criteria for a significant positive effect.

In experiment 2 there was again a significant increase in the incidence of chromosome deletions at the higher dose levels (Table 7.9). There were also significant increases in the incidence of cells with an aberration at all dose levels of 10 ppm and above. The incidence of gaps was significantly greater at 1, 100 and 1000 ppm. The linear trend test showed significant increases of chromatid deletions and exchanges, chromosome deletions, gaps and all types of aberrations with or without gaps. Significant increases in the proportion of abnormal cells were seen with doses of 10 ppm and above, together with highly significant linear trends of most types of aberrations: this gave good evidence of a positive effect.

Further cells were examined from the animals in this study and the results

Table 7.5.1. *Data set D. Chromosome aberration data from animals sampled 6 hours after injection of compound 'Y'.*

		Number of cells with						
	Number of cells		Chromosome		Chromatid		Cells with aberrations	
Treatment	scored	Gaps	Deletions	Exchange	Deletions	Exchange	(+ gaps)	(− gaps)
Negative control	50	2	0	0	2	0	4	2[a]
	50	0	0	0	0	0	0	0
	50	0	0	0	0	0	0	0
	50	0	0	0	1	0	1	1
	50	0	0	0	2	0	3	3[a]
	50	0	0	0	1	0	1	1
	50	0	2	0	0	0	2	2
	50	0	0	0	0	0	0	0
	50	0	0	0	0	0	0	0
	50	0	0	0	0	0	0	0
	50	0	0	0	0	0	0	0
	50	1	1	0	0	0	2	1
40 mg/kg 'Y'	50	1	0	0	0	0	1	0
	50	0	1	0	0	0	1	1
	50	1	0	0	0	0	1	0
	50	1	0	0	1	0	2	1
	50	0	0	0	0	0	0	0
	50	1	1	0	0	0	2	1
	50	0	1	0	0	0	1	1
	50	0	0	0	0	0	0	0

120 mg/kg 'Y'	50	0	0	0	0	0	0	0
	50	0	0	0	1	0	1	1
	50	0	0	0	1	0	1	1
	50	0	0	0	0	0	0	0
	0	—	—	—	—	—	—	—[b]
	50	1	0	0	1	0	2	1
	50	0	0	1	0	0	1	1
	50	0	0	0	0	0	0	0
400 mg/kg 'Y'	50	1	0	0	1	0	2	1
	50	0	1	0	0	0	1	1
	50	0	0	1	0	0	1	1
	50	0	0	0	0	0	0	0
	50	1	0	0	2	0	3	2[a]
	50	0	1	0	0	0	1	1
	50	0	0	0	0	0	0	0
	50	0	0	0	0	0	0	0
Positive control (*EMS*)	50	0	0	0	1	0	1	1
	50	0	1	0	1	0	2	2
	50	0	0	0	1	0	1	1
	50	1	0	0	2	0	3	2
	53	0	0	0	0	0	0	0
	50	2	5	0	0	0	5	4
	50	0	1	0	0	0	1	1
	50	0	0	0	0	0	0	0

Notes:
[a] includes 1 isolocus deletion.
[b] insufficient spreads found.

Table 7.5.2. *Data set D. Chromosome aberration data from animals sampled 24 hours after injection of compound 'Y'.*

Treatment	Number of cells scored	Number of cells with					Cells with aberrations	
		Gaps	Chromosome		Chromatid			
			Deletions	Exchange	Deletions	Exchange	(+ gaps)	(− gaps)
Negative control	50	0	0	1	4	0	5	5[a]
	50	1	0	0	0	0	1	0
	50	0	0	1	0	0	1	1
	52	3	0	0	4	0	6	4
	50	0	1	0	2	0	3	3
	50	1	2	0	1	0	3	3
	50	0	0	0	0	0	0	0
	50	0	0	1	1	0	1	1
	50	0	0	0	1	0	1	1
	50	1	0	0	0	0	1	0
	50	3	3	0	0	0	5	3
	50	0	0	0	0	0	0	0
40 mg/kg 'Y'	50	0	0	0	0	0	0	0
	50	0	0	0	0	0	0	0
	50	0	1	0	0	0	1	1
	50	0	0	0	0	0	0	0
	50	0	0	0	0	0	0	0
	50	1	0	0	0	0	1	0
	50	0	1	0	0	0	1	1
	50	0	0	0	0	0	0	0

120 mg/kg 'Y'	50	0	0	0	0	0	0	0
	50	0	0	0	0	0	0	0
	50	1	1	0	0	0	2	1
	50	0	0	0	0	0	0	0
	50	0	0	0	0	0	0	0
	50	1	0	0	2	0	3	2
	50	0	0	0	0	0	0	0
	50	0	0	0	0	0	0	0
400 mg/kg 'Y'	50	0	0	0	0	0	0	0
	50	0	1	0	1	0	3	3[a]
	50	0	0	0	2	0	2	2
	50	0	1	0	0	0	1	1
	0	—	—	—	—	—	—	—[b]
	50	0	0	0	1	0	1	1
	0	—	—	—	—	—	—	—[b]
	0	—	—	—	—	—	—	—[b]
Positive control (*EMS*)	50	3	4	0	2	0	7	6
	50	0	0	0	1	0	1	1
	50	4	1	0	0	0	4	1
	50	1	1	0	0	0	2	1
	50	1	4	2	7	0	11	11[a]
	52	1	0	2	4	0	6	5
	50	0	0	0	1	0	1	1
	50	2	0	0	1	0	4	2[a]

Note:
[a] includes 1 isolocus deletion.
[b] insufficient spreads found.

Table 7.5.3. *Data set D. Chromosome aberration data from animals sampled 48 hours after injection of compound 'Y'.*

Treatment	Number of cells scored	Gaps	Number of cells with: Chromosome Deletions	Chromosome Exchange	Chromatid Deletions	Chromatid Exchange	Cells with aberrations (+ gaps)	Cells with aberrations (− gaps)
Negative control	50	0	1	0	1	0	2	2
	50	0	3	0	0	0	3	3
	50	0	0	0	0	0	1	1[a]
	50	0	0	0	2	0	2	2
	50	0	0	0	0	0	0	0
	50	0	2	0	0	0	2	2
	50	0	0	0	0	0	0	0
	50	0	0	0	0	0	0	0
	50	1	0	0	1	0	2	1
	50	1	0	0	0	0	1	0
	50	0	0	0	0	0	0	0
	50	0	0	0	1	0	1	1
40 mg/kg 'Y'	0	—	—	—	—	—	—	—[b]
	50	0	0	0	0	0	0	0
	50	1	0	0	0	0	1	0
	50	0	0	0	1	0	1	1
	50	1	0	0	1	0	2	1
	50	1	0	0	0	0	1	0
	50	1	0	0	1	0	2	1
	50	0	0	0	0	0	0	0

120 mg/kg 'Y'	50	0	0	0	0	0	0	0
	50	1	0	0	1	0	2	1[a]
	50	2	0	0	0	0	2	0
	50	2	0	0	0	0	2	0
	50	0	0	0	0	0	0	0
	50	0	1	0	0	0	1	1
	50	1	1	0	1	0	3	2
	50	0	0	0	0	0	0	0
400 mg/kg 'Y'	50	0	0	0	0	0	2	0
	50	1	0	0	0	0	0	0
	50	2	0	0	0	0	2	0
	50	2	0	0	0	0	0	0
	50	0	0	0	0	0	0	0
	50	0	0	0	2	0	2	2[a]
	50	1	0	0	0	0	2	0
	50	0	0	0	0	0	0	0
Positive control (*EMS*)	50	0	0	0	0	0	0	0
	50	1	1	0	2	0	4	4[a]
	50	2	0	0	0	0	3	1[a]
	50	0	0	0	1	0	1	1
	50	1	0	0	0	0	1	0
	50	0	0	0	1	0	1	1
	50	1	0	0	0	0	1	0
	50	0	1	0	0	0	1	1

Note:
[a] includes 1 isolocus deletion.
[b] insufficient spreads found.

Table 7.6.1. *Data set E. Experiment 1. Chromosome aberration data from animals exposed to compounds 'Z' for 2 hours and metaphase spreads fixed 24 hours later.*

			Number of cells with					
			Chromosome		Chromatid		Cells with aberrations	
Treatment	Number of cells scored	Gaps	Deletions	Exchange	Deletions	Exchange	(+ gaps)	(− gaps)
Control	50	3	0	0	0	0	3	0
	50	1	0	0	0	0	1	0
	51	2	0	0	0	0	2	0
	40	2	0	0	0	0	2	0
	50	2	0	0	0	0	2	0
	23	5	0	0	0	0	5	0
	50	2	0	0	0	0	2	0
	50	1	0	0	0	0	1	0
	29	0	0	0	0	0	0	0
	69	4	0	0	1	0	5	1
	27	1	0	0	0	0	1	0
	50	1	0	0	0	0	1	0
1 ppm 'Z'	50	1	0	0	0	0	1	0
	50	3	0	0	0	0	3	0
	35	4	0	0	0	0	4	0
	50	2	0	0	0	0	2	0
	57	2	0	0	0	0	2	0
	50	2	0	0	0	0	2	0
	46	3	0	0	0	0	3	0
	50	4	0	0	0	0	4	0

10 ppm, 'Z'	58	3	0	0	0	0	3	0
	50	1	0	0	0	0	1	0
	55	0	0	0	0	0	0	0
	50	0	0	0	0	0	0	0
	50	3	0	0	0	0	3	0
	50	1	0	0	0	0	1	0
	13	0	0	0	0	0	0	0
	41	4	0	0	0	0	4	0
100 ppm 'Z'	50	0	0	0	0	0	0	0
	50	4	0	0	0	0	4	0
	50	3	0	0	0	0	3	0
	50	2	0	0	0	0	2	0
	50	2	1	0	0	0	3	1
	50	2	2	0	0	0	4	2
	50	5	1	0	0	0	6	1
	50	2	2	0	0	1	3	1
1000 ppm 'Z'	52	3	0	0	0	0	3	0
	50	5	0	0	0	0	5	0
	49	2	0	0	0	0	2	0
	50	3	0	0	0	0	3	0
	50	1	0	0	0	0	1	0
	18	1	2	0	0	0	2	2
	45	5	1	0	0	0	6	1
	14	1	0	0	0	0	1	0

Table 7.6.2. *Data set E. Experiment 2. Chromosome aberration data from animals exposed to compound 'Z' for 6 hours and metaphase spreads fixed 24 hours later*

			Number of cells with					
	Number of cells		Chromosome		Chromatid		Cells with aberrations	
Treatment	scored	Gaps	Deletions	Exchange	Deletions	Exchange	(+ gaps)	(− gaps)
Control	53	0	0	0	0	0	0	0
	56	0	0	0	0	0	0	0
	65	1	0	0	0	0	1	0
	57	2	0	0	0	0	2	0
	59	2	0	0	0	0	2	0
	54	2	0	0	0	0	2	0
	57	0	0	0	0	0	0	0
	55	0	0	0	0	0	0	0
	54	0	0	0	0	0	0	0
	50	1	0	0	0	0	1	0
	53	1	0	0	0	0	1	0
	29	0	0	0	0	0	0	0
1 ppm 'Z'	54	3	0	0	0	0	3	0
	53	2	0	0	1	0	3	1
	60	4	1	0	0	0	5	1
	52	1	0	0	0	0	1	0
	51	0	0	0	0	0	0	0
	60	4	0	0	0	0	4	0
	53	2	0	0	0	0	2	0
	51	1	1	0	0	0	2	1

10 ppm 'Z'	36	1	0	0	1	0	2	1
	51	1	0	0	0	0	1	0
	53	2	1	0	0	0	2	1
	60	1	0	0	0	0	1	0
	36	4	0	0	0	0	4	0
	56	1	2	0	0	1	3	2
	54	0	0	0	0	0	0	0
	59	2	0	0	0	0	2	0
100 ppm 'Z'	50	1	0	0	1	0	2	1
	50	2	1	0	0	0	3	1
	59	4	0	0	0	0	4	0
	36	3	0	0	1	0	4	1
	56	1	0	0	1	0	2	1
	57	2	0	0	0	0	2	0
	51	2	0	0	0	0	2	0
	52	3	0	0	0	0	3	0
1000 ppm 'Z'	56	3	1	0	0	0	4	1
	50	1	1	0	0	1	2	1
	56	3	0	0	0	0	3	0
	51	1	2	0	0	1	3	2
	52	1	1	0	0	0	2	1
	53	3	2	0	1	0	5	3
	50	2	1	0	1	0	3	1
	55	3	2	0	0	0	5	2

Table 7.6.3. *Data set E. Experiment 2. Chromosome aberration data from animals exposed to compound 'Z' for 6 hours and metaphase spreads fix 24 hours later. Results of reading extra cells.*

		Number of cells with						
	Number of cells		Chromosome		Chromatid		Cells with aberrations	
Treatment	scored	Gaps	Deletions	Exchange	Deletions	Exchange	(+ gaps)	(− gaps)
Control	252	4	3	0	1	0	8	4
	256	4	0	0	0	0	4	0
	265	1	0	0	0	0	1	0
	257	4	0	0	0	0	4	0
	259	9	7	0	0	0	16	7
	254	6	1	0	0	0	7	1
	257	4	0	0	0	0	4	0
	255	0	0	0	0	0	0	0
	254	1	0	0	0	0	1	0
	50	1	0	0	0	0	1	0
	253	3	0	0	0	0	3	0
	29	0	0	0	0	0	0	0
1 ppm 'Z'	266	5	0	0	0	0	5	0
	253	4	0	0	1	0	5	1
	260	7	1	0	0	0	8	1
	252	5	2	0	0	0	7	2
	251	5	0	0	0	0	5	0
	60	4	0	0	0	0	4	0
	253	4	3	0	0	0	7	3
	251	1	1	0	0	0	2	1

10 ppm ‘Z’	36	1	0	0	1	0	2	1
	239	3	3	0	0	0	6	3
	253	2	1	0	0	0	3	1
	260	2	0	0	0	0	2	0
	36	4	0	0	0	0	4	0
	56	1	2	0	0	1	3	2
	54	0	0	0	0	0	0	0
	259	7	0	0	0	0	7	0
100 ppm ‘Z’	250	5	0	0	1	0	6	1
	68	2	1	0	0	0	3	1
	259	9	4	0	0	0	13	4
	36	3	0	0	1	0	4	1
	56	1	0	0	1	0	2	1
	257	14	4	0	0	0	18	4
	251	7	0	0	0	0	7	0
	252	4	0	0	0	0	4	0
1000 ppm ‘Z’	256	9	8	0	1	0	18	9
	266	5	2	0	0	1	8	3
	256	6	5	0	0	0	11	5
	251	1	16	0	0	1	18	17
	252	9	10	0	0	0	19	10
	253	8	16	0	1	0	25	17
	250	6	10	0	1	0	17	11
	255	5	9	0	0	0	14	9

Table 7.7. *Heterogeneity chi-squares for differences between animals within groups in chromosome aberration data set D.*

Sampling time (h)	Dose mg/kg	Total cells scored	Total cells with aberration (excluding gaps)		Total cells with aberration (including gaps)	
6	0	600	10	14.24[a]	13	19.74*
	40	400	4	4.04	8	4.08
	120	350	4	3.03	5	4.87
	400	400	6	4.74	8	8.16
24	0	602	21	20.05*	27	21.97*
	40	400	2	6.03	3	5.04
	120	400	3	10.41	5	16.00*
	400	250	7	3.82	7	3.82
48	0	600	12	12.24	14	10.24
	40	350	3	4.03	7	4.08
	120	400	4	8.08	10	7.79
	400	400	2	14.07*	8	8.16

Notes:
[a] heterogeneity χ^2 for variation between animals within groups.
* $P<0.05$ (from a table of the χ^2 distribution).

obtained are shown in Table 7.10. Highly significant increases in the incidence of chromosomal deletions were again found together with increases in the proportion of cells showing abnormalities including or excluding gaps at the 1000 ppm dosage. Only an increase in gaps was significant at any of the lower doses (100 ppm). This contradiction in the results between the two parts of experiment 2 is probably related to the heterogeneity in the control rats which had a higher incidence of chromosome deletions in the second part of the study. Much of the heterogeneity could be traced to one animal (No. 5) which had 7 out of 259 cells with chromosomal deletions. It is an interesting question whether to include or exclude such an animal in the analysis. Statistical reanalysis with the potential outlier excluded is one course of action.

The extra cells examined in experiment 2 were scored with the purpose of trying to clarify the dose-reponse relationship seen with this compound. Whether it is more cost-effective to score more cells in fewer animals than outlined in the present guidelines depends very much upon the degree of variability or heterogeneity observed among the animals used in these assays. An analysis of the results of a series of studies could help to answer these questions.

Box 7.9. *2 × 2 chi-square tests and linear trend tests of chromosome aberration data set D.*

	Test	Sampling time 6 h	24 h	48 h
Proportion of cells with aberrations excluding gaps	0 v. 40	−0.604[a]	−2.878	−1.093
	0 v.120	−0.367	−2.565	−0.977
	0 v.400	0.051	−0.302	−1.703
	Linear trend	0.066[b]	−0.047	−1.788
Proportion of cells with aberrations including gaps	0 v. 40	0.045	−3.208	−0.108
	0 v.120	−0.558	−2.669	−0.042
	0 v.400	0.045	−0.952	−0.132
	Linear trend	−0.128	−0.752	−0.279

Notes:

[a] z value obtained from comparison of control and treated proportion of abnormal cells. Calculation based upon χ_c^2 test with continuity correction (see Box 7.7) where $z = \sqrt{(\chi_c^2)}$. Positive or negative sign determines whether control proportion has lower or higher proportion of abnormal cells than exposed. Note that where the incidence is similar in the two groups a positive value of z may be obtained even when the control proportion is higher than the exposed. This is because of the correction for continuity.

An alternative formulation for the test can be found in Snedecor & Cochran (1968, p. 220).

Values of z are tested using tables of the normal distribution.

[b] Linear trend test value (z) of increasing incidence of aberrations with increasing dose (calculated as in Box 7.8) where dose is 0, 40, 120 and 400 mg/kg.

7.3.2 The rodent germ cell (spermatogonial) cytogenetic assay

One direct test of chromosomal damage in the germ cells of dosed animals is the spermatogonial cytogenetic assay. Metaphase spreads from spermatogonial cells of dosed animals are analysed for chromosomal aberrations in a similar way to the bone marrow cytogenetic studies. The statistical methods for analysing the data are similar.

The statistical analysis of data from other direct tests such as meiotic figures from spermatocytes or indirect tests such as the heritable translocation test have not been considered here. Full details of these methods are given in Albanese *et al.* (1984).

Table 7.8. *Tests of aberration data from chromosomal aberration data set E. Experiment 1. Acute 2-hour exposure 24-hour kill.*

		Number of cells with abnormalities					
		Chromatid					
Dose (ppm)	Total cells scored	Deletions	Exchanges	Chromosome deletions	Gaps	Total (without gaps)	Total (with gaps)
0	539	1[a]	0	0	24	1	25
1	388	0	0	0	21	0	21
		0.165[b]	—	—	0.516	0.165	0.382
10	367	0	0	0	12	0	12
		0.193	—	—	−0.722	0.193	−0.851
100	400	0	1	6	20	5	25
		0.150	0.150	2.438**	0.236	1.610	0.941
1000	328	0	0	3	21	3	23
		0.251	—	1.628	1.097	1.020	1.329
Linear trend		−1.251[c]	0.837	3.215***	0.928	2.517**	1.493

Notes:

[a] number of cells showing particular aberrations

[b] z value of test of proportion of aberrant cells in treated group compared with control group.

[c] z value of test for linear trend for increasing incidence of aberrations with increasing dose (based upon doses (D_i) coded 0, 1, 2, 3, 4).

Significance levels (one-sided tests) *$P<0.05$, **$P<0.01$, ***$P<0.001$ of tests of values of z using tables of normal distribution.

Table 7.9. *Tests of aberration data from chromosomal aberration data set E. Experiment 2. Acute 6-hour exposure 24-hour kill*

Dose (ppm)	Total cells scored	Number of cells with abnormalities					
		Chromatid					
		Deletions	Exchanges	Chromosome deletions	Gaps	Total (without gaps)	Total (with gaps)
0	642	0[a]	0	0	9	9	0
1	434	1	0	2	17	20	3
		0.197[b]	—	1.000	2.433**	2.994**	1.520
10	405	1	1	3	12	15	4
		0.233	0.233	1.590	1.528	2.212*	2.009*
100	411	3	0	1	18	22	4
		1.575	—	0.225	2.782**	3.513***	1.991*
1000	423	2	2	10	17	27	11
		1.021	1.021	3.590***	2.505**	4.228***	3.798***
Linear trend		1.902*[c]	1.805*	3.782***	2.549**	4.065***	3.986***

Notes:
[a] number of cells showing particular aberrations
[b] z value of test of proportion of aberrant cells in treated group compared with control group.
[c] z value of test for linear trend for increasing incidence of aberrations with increasing dose.
Significance levels (one-sided tests) $^{*}P<0.05$, $^{**}P<0.01$, $^{***}P<0.001$ of tests of values of z using tables of normal distribution.

Table 7.10. *Tests of aberration data from chromosomal aberration data set* E. *Experiment 2 (extra data). Acute 6-hour exposure 24-hour kill re-evaluation with extra cells.*

Dose (ppm)	Total cells scored	Number of cells with abnormalities					
		Chromatid					
		Deletions	Exchanges	Chromosome deletions	Gaps	Total (without gaps)	Total (with gaps)
0	2641	1[a]	0	11	37	12	49
1	1846	1	0	7	35	8	43
		−0.464[b]	—	0.045	1.178	0.124	0.995
10	1193	1	1	6	20	7	27
		−0.187	0.408	0.110	0.508	0.292	0.714
100	1429	3	0	9	45	12	57
		1.148	—	0.694	3.672***	1.318	3.976***
1000	2039	3	2	76	49	81	130
		0.764	0.897	8.205***	2.421**	8.446***	7.918***
Linear trend		1.644[c]	1.699*	9.465***	3.220***	9.739***	8.690***

Notes:
[a] number of cells showing particular aberrations
[b] z value of test of proportion of aberrant cells in treated group compared with control group.
[c] z value of test for linear trend for increasing incidence of aberrations with increasing dose.
Significance levels (one-sided tests) *$P<0.05$, **$P<0.01$, ***$P<0.001$ of tests of values of z using tables of normal distribution.

7.3.3 Discussion of the various approaches to the analysis of chromosomal aberration data

The preferred method for the analysis of chromosomal aberration data is the use of χ^2 tests. An alternative approach, using the analysis of variance after arcsin transformation, is possible. This transformation changes data which are binomially distributed into a form which is approximately normally distributed. Although superficially different, the χ^2 and analysis of variance after arcsin transformations are very similar, and in fact are special cases of the generalised linear model referred to earlier. Both methods can be analysed in a number of ways using either the Genstat or GLIM packages. The χ^2 test is easier to use since the analysis of variance (using the arcsin) involves a theoretical residual variance based on the binomial distribution, the derivation of which is not immediately apparent to a non-statistician.

There is some debate on whether to pool scores over animals within a group or to test for heterogeneity within the groups. If heterogeneity is found it has been suggested that modified statistical tests should be carried out. The compromise approach suggested here is that a test is carried out for heterogeneity, the results are reported but χ^2 tests are carried out regardless. Any heterogeneity within the groups would then be one of the factors included in the interpretation of the data.

7.4 GENERAL DISCUSSION OF ISSUES ARISING DURING *IN VIVO* CYTOGENETIC STUDIES

Irrespective of the statistical results there are some biological principles which should be kept in mind during the conduct and analysis of *in vivo* cytogenetic and micronucleus experiments. They relate to potential variations in the yield of aberrations over time and the nature of the mitotic index.

7.4.1 A cautionary note on yield–time variation

Great care is needed if quantitative comparisons are made between different treatments or substances. Two agents producing potentially the same levels of aberration may perturb populations of cells in different ways. This can result in quite different 'yield–time' profiles of micronuclei or chromosomal aberrations.

Both chromosomal aberration yield and the induction of micronuclei will consequently vary over time. This variation will increase as a result of different degrees of mitotic delay and 'disturbances' produced by a clastogen acting both within a population of cells and between populations receiving different doses.

Cells also have different sensitivity for structural chromosomal aberrations at different times in the cell cycle so that in an asynchronous population there is no unique aberration frequency for a specific dose (Savage, 1978).

The profile of micronucleus frequency with time therefore bears no simple relationship to aberration induction. Attempts to deduce mechanism of action by mathematical analysis of dose–response curves using measures such as peak/plateau yield, the total integrated area under the yield–time curve or the overall mean will be unsuccessful.

7.4.2 A note on 'mitotic index'

The Mitotic Index (MI) is defined as 'the ratio' of the number of cells seen in mitosis to the total number of cells present. Only in an ideal 'steady-state' population is there a simple relationship between MI and mitotic duration. Such populations are seldom realised especially when the cell population is disturbed by clastogenic treatments. Simple interpretation of MI is therefore not always possible and MI is not therefore a useful measure of 'mitotic rate'. Treatments, for instance, which affect the differential length of interphase and mitosis can result in either an increase or decrease in MI.

7.5 DECIDING WHETHER A RESULT IS POSITIVE OR NEGATIVE

The choice of statistical analysis is only part of the process for deciding whether the test chemical is classified as positive or negative in the test. This assignment can be made either on the basis of a precise decision rule or on a more subjective assessment of the evidence.

In either case there are two areas where the decision may be wrong. First, the decision rule may produce a false positive or negative. This will be a function of the experimental design, the statistical tests used and the rule chosen. Estimating the actual error rates of different rules is a complex area but one worthy of further work. Secondly, the model system may only be a partial predictor of human risk. Such errors in classification of a substance are in general beyond the scope of the discussion here.

Decision rules are implicit in any guidelines for cytogenetic tests. The suggestion of Scott *et al.* (1983) can be summarised as follows: a compound produced a positive response if two or more dose levels were significantly increased over the negative control, these increases were dose related and

Table 7.11. *Decision rule used by Galloway* et al. *(1985) in their analysis of* in vitro *chromosomal aberration data. Experimental design: 100 cells scored at each of five doses for chromosomal aberrations.*

Number of significant differences between control and dose groups (2×2 tests) at $P<0.01$	Trend test significant at		
	$P>0.005$	$0.005>P>0.001$	$0.001>P$
0	–	–	?w
1	–	?w	?+
2	+b	+	+

Note:
Where – is negative; + is positive ?w and ?+ are weak positive and questionable positive and +b is positive but lacking in dose–response.
Source: (from Galloway *et al.*, 1985).

the responses were greater than a predefined size depending upon knowledge of historical data. The damage seen should also be consistent with that expected, taking into account the time of sampling.

In some cases, as with micronucleus data, the dose related response is referred to as a 'biologically sensible relationship'.

Such decision rules are somewhat subjective: the increase over background is arbitrary, while the significance levels will depend upon the power of the statistical tests used.

Galloway *et al.* (1985) provide an illustration of a more explicit decision rule which incorporates the results from statistical tests in their assessment of results of *in vitro* chromosomal aberrations.

Table 7.11 shows the decision rule used. Chemicals were assigned to one of five categories: negative, positive with dose-related trend, positive with lack of dose–response relationship, possible positive and possible weak positive. Similar decision rules could no doubt be devised for the combination of experimental designs and statistical tests in the *in vivo* tests considered here. Further work on the development of such decision rules should be considered.

Guidelines for carrying out *in vivo* cytogenetic studies propose that the statistical methods and criteria for defining positive results should be established before a study is started. This objective should not be rigidly adhered to as a consensus of absolute rules is unlikely and probably undesirable. A blend of good biology, good cytogenetics and good statistics is more likely to lead to a correct assessment of genotoxicity.

7.6 CONCLUSIONS

(a) The aim of any statistical analysis should be to provide the optimum description of the data available to help in deciding future action on the test chemical.

(b) Micronucleus assay data can be satisfactorily analysed by a range of statistical methods including analysis of variance after transformation, general linear modelling, chi-square tests, likelihood ratio tests, Kruskal–Wallis and Mann–Whitney non-parametric tests.

(c) The choice of the test will be based upon the personal preference of the individual laboratory and will, in part, reflect the facilities available to them.

(d) Chromosomal aberration data are best analysed using χ^2 tests although tests based on the analysis of variance are also satisfactory.

(e) Tests of heterogeneity within groups should be reported but it should be left to the investigator to decide what further action is needed on the analysis of the data.

(f) Tests for linear trend with dose should be included where appropriate because these are likely to be the most sensitive statistical tests.

(g) Decision rules should be defined more specifically to try to help the interpretation of the data collected in these studies.

(h) Further work should be undertaken to investigate the relative powers of different experimental designs and statistical tests. Research aimed at estimating the error rates associated with particular decision rules should be carried out.

7.7 REFERENCES

Albanese, R., Topham, J., Evans, E., Clare, G. and Tease, C. (1984). Mammalian germ cell cytogenetics in: *UKEMS Sub-committee on Guidelines for Mutagenicity Testing,* Report, Part II, Supplementary Tests, Ed. B.J. Dean, United Kingdom Environmental Mutagen Society, Swansea, pp.145–72.

Alvey, N.G., Galwey, N. & Lane, P. (1982). *An Introduction to Genstat,* Academic Press, London.

Amphlett, G.E. & Delow, G.F. (1984). Statistical analysis of the micronucleus test. *Mutation Research,* **128**, 161–6.

Ashby, J. and Mohammed, R. (1986). Slide preparation and sampling as a major source of variability in the mouse micronucleus assay. *Mutation Research,* **164**, 217–35.

Baker, R.J. & Nelder, J.A. (1978). The GLIM System Manual Release 3. Numerical Algorithms Group, Oxford.

Brøgger, A. (1982). The chromatid gap – a useful parameter in genotoxicity. *Cytogenetics and Cell Genetics*, **33**, 14–19.
Campbell, R.C. (1974). *Statistics for Biologists*. 2nd Edn. Cambridge University Press, Cambridge.
Couzin, D. & Papworth, D.G. (1979). The over-dispersion between cells of chromosomal aberrations. *Journal of Theoretical Biology*, **80**, 249–58.
Department of Health and Social Security (1981). Guidelines for the testing of chemicals for mutagenicity. Report on Health and Social Subjects 24, *Her Majesty's Stationery Office*, London.
Everitt, B.S. (1977). *The Analysis of Contingency Tables*. Chapman & Hall, London.
Galloway, S.M., Bloom, A.D., Resnick, M. Margolin, B.H., Nakamura, F., Archer, P. & Zeiger, E. (1985). Development of a standard protocol for *in vitro* cytogenetic testing with chinese hamster ovary cells: comparison of results for 22 compounds in two laboratories. *Environmental Mutagenesis*, **7**, 1–51.
Heddle, J.A., Hite, M., Kirkhart, B., Mavournin, K., MacGregor, J.T., Newell, G.W. & Salamore, M.F. (1983). The induction of micronuclei as a measure of genotoxicity. A report of the US Environmental Protection Agency Gene-Tox Program. *Mutation Research*, **123**, 61–118.
Kastenbaum, M.A. & Bowman, K.O. (1970). Tables for determining the statistical significance of mutation frequencies. *Mutation Research*, **9**, 527–49.
Mackey, B.E. & MacGregor, J.T. (1979). The micronucleus tests: statistical design and analysis. *Mutation Research*, **64**, 195–204.
Margolin, B.H., Collings, B.J. & Mason, J.M. (1983). Statistical analysis and sample-size determination for mutagenicity experiments with binomial response. *Environmental Mutagenesis*, **5**, 705–16.
Margolin, B.H., Resnick, M.A., Rimpo, J.Y., Archer, P., Galloway, S.M., Bloom, A.D. & Zeiger, E. (1986). Statistical analysis for *in vitro* cytogenetic assays using chinese hamster ovary cells. *Environmental Mutagenesis*, **8**, 183–204.
Mitchell, I. de G., Dixon, P.A. & White, D.J. (1981). Analysis of *in vivo* results of cyclophosphamide-induced chromosomal damage in mammals from sensitivity and statistical aspects. *Journal of Toxicology and Environmental Health*, **7**, 585–92.
Mitchell, I. de G. & Brice, A.J. (1986). Investigations into parametric analysis of data from *in vivo* micronucleus assays by comparison with non-parametric methods. *Mutation Research*, **159**, 139–46.
Papworth, D.G. (1983). Exact tests of fit for a Poisson distribution. *Computing*, **31**, 33–45.
PSI (1986). Report of PSI working party on the data produced by the ABPI collaborative study into the micronucleus test.
Savage, J.R.K. (1978). Some thoughts on the nature of chromosomal aberrations and their use as a quantitative end-point for radiobiological studies. pp. 39–54 in *Sixth Symposium on Microdosimetry* I (ed. J. Booz, H.G. Ebert) C.E.C. Harwood, Belgium.
Scott, D., Danford, N., Dean, B., Kirkland, D. & Richardson, C. (1983). *In vitro* chromosome aberration assays, in *UKEMS Sub-Committee on Guidelines for Mutagenicity Testing*, Report, Part I, Basic Test Battery, Ed. B.J. Dean, United Kingdom Environmental Mutagen Society, Swansea, pp. 41–64.
Snedecor, G.W. & Cochran, W.G. (1968). *Statistical Methods* 6th Edn, Iowa State University Press, Iowa.

Snedecor, G.W. & Cochran, W.G. (1980). *Statistical Methods* 7th Edn, Iowa State University Press, Iowa.

Topham, J., Albanese, R., Bootman, J., Scott, D. & Tweats, D. (1983). *In vivo* cytogenetic assays, in: *UKEMS Sub-committee on Guidelines for Mutagenicity Testing*, Report, Part I, Basic Test Battery, Ed. B.J. Dean, United Kingdom Environmental Mutagen Society, Swansea, pp. 119–141.

Yates, F. (1934). Contingency tables involving small numbers and the χ^2 test. *Journal of the Royal Statistical Society Supplement*, **1**, 217–35.

8

Statistical methods for the dominant lethal assay

D.O. CHANTER D. ANDERSON
A. BATEMAN D.P. LOVELL
A.K. PALMER M.T. STEVENS
M. RICHOLD (Group Leader)

8.1 INTRODUCTION

The dominant lethal assay, conducted in mice or rats, is used to detect agents which produce genetic abnormalities in sperm capable of affecting the viability of the foetus. In its usual form, the males are treated acutely (or sub-acutely) with the test compound, and then mated with one or more untreated virgin females once a week for a complete spermatogenesis cycle (8 weeks in mice and 10 weeks in rats). At about 15 days after mating the females are killed and their uteri examined. The number of live implants, and the number of early and late deaths, are recorded. Fuller details of the experimental procedures are given by Anderson *et al.* (1983).

In this chapter, we confine our attention to the dominant lethal assay as outlined above; variants of the assay in which the male is treated for several weeks with only one mating, or in which the females are treated, are not considered.

The data generated by a dominant lethal assay are statistically quite complex: several different sources of variation are present, including the treatment effect itself, variation between males, variation between females and inherent 'statistical' variation. It is therefore not surprising that a large number of methods of statistical analysis have been suggested.

It is doubtful whether any single method will ever become generally accepted as *the* correct way to analyse data from a dominant lethal assay; some methods can clearly be identified as inappropriate, but the relative merits of the remainder are likely to continue to be a source of argument for a long time to come.

Because of the complexity of the data themselves, many of the suggested techniques are statistically quite complex. However, some are relatively straightforward and it is these we concentrate on in this chapter. The more complex methods can be used by workers with professional statistical help available. The gain in power (see Chapter 1) to be obtained with these more complex methods depends on the design of the study and the relative magnitudes of the various sources of variation, and in many cases only a small increase in power is likely to be achieved. However, in comparison with the cost of carrying out the study the cost of professional statistical assistance with the analysis will also be small.

8.2 PRELIMINARY CONSIDERATIONS

In the following sections we consider various aspects of the assay which impinge on the statistical analysis, and which need to be considered before the analysis can begin.

8.2.1 Design

In order for the statistical analysis of the experimental data to be valid, it is necessary that the study be designed according to certain principles. One of the most important of these is randomisation, and to be effective this must be carried out deliberately, e.g. using random number tables, rather than relying on a haphazard method of allocating animals to treatment groups.

The simplest form of randomisation is *full randomisation*. With this scheme, if for example there were 100 animals to be assigned to 4 groups of 25 each, the animals would be numbered 1 to 100 (in any order), and a random permutation of the numbers 1 to 100 obtained. The animals corresponding to the first 25 numbers in the permutation would then form Group 1, the next 25 Group 2, and so on.

The main drawback of full randomisation is that it can, by chance, lead to a batch of similar animals all being assigned to the same group: for example, if the 100 animals contained some litter mates, there is nothing to prevent these ending up in the same treatment group. This problem can be avoided by the use of *blocking*. To apply blocking in the example used above, the 100 animals would first be divided into 25 *blocks* of 4 animals each. Within each block the animals would be as similar as possible (i.e. litter mates, or of similar bodyweight). Using a different random permutation of the numbers 1 to 4 for each block, the four animals in any block are assigned one to each of the treatment groups.

For the dominant lethal assay, the randomisation procedure has to be repeated several times; once for the males, and again for each group of females used. If some of the females to be used come from the same litters as some of the males, the randomisation procedure should be altered to avoid pairings between litter mates.

The use of a restricted randomisation procedure such as blocking will be effective in making the study more powerful only if the variables used for blocking (litter, bodyweight, etc.) are actually correlated with the response variables to be analysed. However, if they are not, the use of blocking is unlikely to weaken the experiment much.

When a study has been structured in blocks, it is important that this fact should be taken into account in the statistical analysis, otherwise the effort will have been largely wasted (see Box 8.6). Further discussion on blocking, and on other features of design, in the general context of toxicology studies, is given by Healey (1983).

Often in dominant lethal assay studies a positive control group is included. The function of this group is to demonstrate that the assay is sensitive to the effect of a known positive agent. If the effect in this group is so large as to be apparent without statistical analysis, this group is best excluded from the analysis of the remaining groups; this is because it is likely to have a larger variance than the other groups, and its inclusion could reduce the sensitivity of the statistical analysis.

8.2.2 Choice of variables for analysis

In a standard dominant lethal assay, there are a number of measurable variables that may be considered for analysis. The following list, though not exhaustive, contains those directly measured variables commonly considered:

(a) Clinical signs.
(b) Change in bodyweight of the males.
(c) Number of deaths among the males.
(d) Number of pregnancies.
(e) Number of corpora lutea.
(f) Total number of implants (*TI*).
(g) Number of early deaths (*ED*) or moles.
(h) Number of late deaths (*LD*).
(i) Number of viable implants (*LI*).

In addition to these directly measured variables a number of derived variables can be calculated (a list of some of these is given by Bateman (1984)). The most commonly chosen derived variable seems to be the number of early deaths per implant (*ED*/*TI*). The reason for excluding the number of late deaths from the numerator is that these deaths might not be the result of dominant lethal processes. It is recommended, however, that the total number of deaths per implant, $(ED+LD)/TI$, is also analysed. Although many other derived variables can be defined, the reader is warned against analysing too many such variables, since this increases the chance of obtaining false positive results.

Previously, some authors (e.g. Dean & Johnstone, 1977) have used the number of early deaths (ED) as the main end-point. However, Lovell *et al.* (1987) have shown that the ratio ED/TI is more sensitive, and therefore this ratio will be used in this chapter to illustrate the statistical procedures. In practice, it is usual to select several different end-points for analysis.

8.2.3 Experimental unit

As pointed out previously (Anderson *et al.*, 1983), there has been some debate as to whether it is appropriate to consider the male or the female as the basic experimental unit. When 1:1 matings are used, the problem does not arise, because the two assumptions lead to the same analysis. However, with 1:2 matings it is necessary to make an appropriate choice for the experimental unit for the methods of analysis considered in this paper. In fact, for designs in which only the males are treated (the only type of design considered in this chapter), there can be no doubt that the appropriate experimental unit is the male, since this is the unit to which the treatment is applied. This means that for the simpler methods of analysis it is necessary to average the data for the two (or more) females mated to the same male in any time period.

For variables which are ratios, such as ED/TI, there is more than one way of carrying out this averaging process, so unfortunately the fact that the male must be used as the basic experimental unit leads to a problem in deciding how to average the data from the females. For example, suppose that for a given case one female has no early deaths out of 12 implants and the other female has one early death out of 5. One method of averaging is to consider this as one early death out of 17 implants; this is effectively equivalent to considering each foetus as contributing an equal amount of information, and gives a value of 1/17 or 0.0588 for the proportion of deaths. Another method of averaging is to work out the ratio separately for each female (0 and 0.2 in this example) and then average these ratios; this is equivalent to considering each female as providing an equal amount of information, and in this case the result is 0.1. Other methods of averaging can be defined, which mostly give answers in between these two results.

The best method of averaging to use in any particular experiment actually depends in a rather complex way on the variability between females mated to each male. In the discussion below. we have used the first method given above for illustrative purposes, but in practice it would be safer to carry out the analysis using both methods. If the same conclusions are reached regardless of which method of averaging is used, then clearly the choice of method is of little importance, but if different conclusions are reached then it would be necessary to have a more detailed analysis carried out by a qualified statistician, who would either identify the appropriate averaging process or use a method of analysis which avoided the problem.

The problem of having to average data from the females is one of the main disadvantages of using mating ratios other than 1:1, and this provides a strong reason for preferring designs which use a 1:1 mating ratio.

8.3 THE STATISTICAL ANALYSIS

We now consider some of the methods available for the analysis of the data. These fall broadly into two categories, non-parametric and parametric. Although the non-parametric approaches rest on fewer assumptions than the parametric approaches, both make the assumption of independence between sample units, which is one reason why proper randomisation at the start of the experiment is important. The parametric approaches can be further subdivided into those based on the normal distribution and those based on the binomial distribution.

8.3.1 Chi-squared test

One simple statistical method which has been suggested for use with these data is the chi-squared test for contingency tables (Bateman, 1984). This method is valid for those variables, such as the number of deaths among the males, in which each male is scored once only as one of two possible outcomes. However, as pointed out by Bateman, this test is not applicable for variables such as the proportion of early deaths if heterogeneity is present between the males. This is because the assumption of independence between implants does not hold in this situation. Since such heterogeneity is so often present, it is our view that this approach should be used only with extreme caution (see also Haseman & Soares, 1976). For similar reasons the Fisher 'exact' test would also not be appropriate.

8.3.2 Non-parametric approaches

A simple non-parametric approach which can be validly applied is the Mann–Whitney *U*-test (sometimes called the Wilcoxon Rank Sum Test). This test, however, is designed to compare only two groups and therefore needs to be separately applied for each pair of groups (using data

Table 8.1. *Data used for demonstration purposes.* (The data shown are the number of early deaths expressed as a proportion of the total number of implants.)

Control			Low			Intermediate			High		
M	F_1	F_2	*M*	F_1	F_2	*M*	F_1	F_2	*M*	F_1	F_2
1	0/11	0/12	16	0/ 0	0/12	31	1/10	0/16	46	0/ 0	0/ 0
2	1/14	1/13	17	1/13	0/11	32	0/13	0/ 9	47	1/12	1/10
3	0/11	0/ 0	18	0/10	0/13	33	0/14	0/ 7	48	1/10	0/ 9
4	1/10	0/ 0	19	0/10	0/10	34	0/ 0	0/12	49	2/10	1/16
5	0/ 0	0/12	20	0/12	0/11	35	0/13	1/14	50	0/ 0	0/11
6	3/11	0/10	21	0/13	1/11	36	0/10	1/11	51	0/ 0	1/11
7	0/10	0/13	22	0/ 0	0/11	37	0/ 5	1/11	52	1/11	1/11
8	0/11	0/11	23	0/11	1/11	38	0/13	0/13	53	0/13	1/15
9	0/ 9	0/ 0	24	1/10	0/11	39	0/ 0	0/ 0	54	0/ 0	0/12
10	0/ 0	0/13	25	0/ 9	1/10	40	2/14	0/11	55	0/ 8	0/12
11	0/15	0/ 0	26	2/15	0/11	41	0/14	1/ 8	56	0/ 9	1/14
12	0/14	3/13	27	1/11	1/12	42	1/12	0/ 0	57	5/ 9	0/13
13	0/10	1/10	28	0/13	0/15	43	0/13	0/10	58	0/10	1/12
14	0/11	0/ 0	29	1/12	0/11	44	1/13	0/13	59	0/12	2/14
15	0/14	0/12	30	1/10	0/14	45	1/12	3/12	60	0/13	1/14

Notes:
M, male number. F_1, first female. F_2, second female.

for only one week) between which a comparison is required. The whole process must then be repeated for each week. The application of the test is illustrated in Boxes 8.1–8.4, using some data given in Table 8.1.

The main problem with using the Mann–Whitney *U*-test routinely is the large number of tests which are carried out; e.g. in a four-group study there are six inter-group comparisons, although perhaps only three of these (those involving the control) would be considered of interest. If the males are mated for each of eight weeks that gives 48 separate comparisons, 24 of which involve the control group. If the test compound has no effect, each one of these comparisons will have a 5% chance of giving a statistically significant result at the 5% level (i.e. the false positive error rate). With 24 comparisons, the chance of at *least one* such result is about 71% (rising to about 92% for 48 comparisons). Thus most 'negative' compounds would be expected to give at least one statistically significant effect.

Most statistical techniques can be modified to take account of this problem; such modifed techniques are known as *multiple comparison procedures* and their aim is generally to reduce the *overall* chance of a significant result (when there is no real difference between treatments) to 5% (or whatever level is chosen). The price that has to be paid for reducing

Box 8.1. *The Mann–Whitney* U*-test; ranking the data*
To illustrate this test we use the data from the Control and Intermediate dose groups shown in Table 8.1. The data are first averaged across the two females mated to each male (see text: here we use only the first method described). Next the males are ranked in order of increasing response (note that Male No. 39 is omitted from the ranking because no usable data are available). Where ties are present (as in the case of zero, which occurs 15 times), the mid-rank is used. This is the average of the ranks which would have been assigned had the data not been tied (e.g. 8 is the average of the numbers 1 to 15). The resulting rankings are shown below:

Control		Intermediate	
Data	Rank	Data	Rank
0/23=0.000	8	1/26=0.038	17.5
2/27=0.074	23	0/22=0.000	8
0/11=0.000	8	0/21=0.000	8
1/10=0.100	26	0/12=0.000	8
0/12=0.000	8	1/27=0.037	16
3/21=0.143	28	1/21=0.048	20
0/23=0.000	8	1/16=0.063	22
0/22=0.000	8	0/26=0.000	8
0/ 9=0.000	8	0/ 0 —	—
0/13=0.000	8	2/25=0.080	24
0/15=0.000	8	1/22=0.045	19
3/27=0.111	27	1/12=0.083	25
1/20=0.050	21	0/23=0.000	8
0/11=0.000	8	1/26=0.038	17.5
0/26=0.000	8	4/24=0.167	29
Sum	205	Sum	230

Notation:
Number of males for which data are available:
Control group: n_1 Treated group: n_3 $n_1+n_3=N$
Sum of ranks:
Control group: R_1 Treated group: R_3
In this case:
$n_1=15$ $n_3=14$ $N=29$ $R_1=205$ $R_3=230$
Calculation check:
R_1+R_3 should be equal to one half of $N(N+1)$

the false positive error rate in this way is of course that the false negative error rate increases. With an assay like the dominant lethal, where the effect of a compound might operate at only one specific part of the spermatogenesis cycle, and thus be apparent in only one or two of the weeks of the study, it is not in our view desirable to reduce the sensitivity of the study by using multiple comparison techniques indiscriminately. Thus we recommend

Box 8.2. *The Mann–Whitney* U-*test; calculation of* U.
Using the notation from Box 8.1 calculate

$$U_{13} = n_1 n_3 + [n_1(n_1+1)]/2 - R_1$$

and

$$U_{31} = n_1 n_3 + [n_3(n_3+1)]/2 - R_3$$

Either of U_{13} and U_{31} can be regarded as the Mann–Whitney statistic, U. Which one is selected depends on which method of determining significance is used (see Box 8.3) and whether a one- or two-tailed test is required. If U_{13} is greater than U_{31} the treated group has a higher incidence of affected foetuses than the control group.

In this case: $U_{13} = 125$ $U_{31} = 85$

Calculation check: $U_{13} + U_{31}$ should be equal to $n_1 n_3$

Box 8.3. *The Mann–Whitney* U-*test; determining the significance of* U.
In cases where the larger of the two group sizes is more than 20, the statistical significance of U is determined by transforming it to a standard normal variate (usually called z) and using tables of the normal distribution. The relevant formulae follow (with numerical values for the example given for illustration, although in this case both group sizes are less than 20 and the results will therefore be somewhat unreliable):

$$z = (U - m)/s$$

where

$m = n_1 n_3/2$ (equals 105.0 in this example)
$s^2 = n_1 n_3 (N+1)/12$ (equals 525 in this example and hence s is equal to 22.91)

Thus in the example, if U_{13} is used, z is equal to 0.873. For significance at the 5% level z needs to be at least 1.64 for a one-tailed test or 1.96 for a two-tailed test. Thus this result is not statistically significant. To use this method of determining significance, use U_{13} for a one-tailed test or the larger of U_{13} and U_{31} for a two-tailed test.

When both sample sizes are less than 20 (as in the example), more accurate assessments of significance can be made (in the absence of extensive ties) from published tables of critical values of U (see e.g. Table K of Siegel, 1956). To use Siegel's tables, the *smaller* of U_{13} and U_{31} is chosen as the Mann–Whitney statistic. Ignoring the ties present in the example used here, the tabulated critical value for U for a study of this size and a one-tailed test at the 5% level is 66, and U_{31} would need to be equal to or less than this value in order to be significant. With this assay, however, extensive ties often are present, so the method of converting to a z-value is usually to be preferred. A modification is available which makes an adjustment for ties (see Box 8.4).

Box 8.4. ***Dealing with ties.***

When ranking the data, ties are dealt with as described in Box 8.1. When calculating the z value in the presence of ties, a different formula for s is required.

First, determine how many ties there are, and how many observations are involved in each tie. Only ties which involve both groups need to be counted. In the example, there is only one tie (on zero) which counts; this is of size (or length) 15. The tie of size 2 (on 0.038) does not count because both observations are in the same group.

The length of each tie is denoted by t, and for each value of t calculate $T=(t^3-t)/12$, and sum the resulting values, calling the result S_t. In the example, S_t, (there is only one T value) is 280. The formula for s is now

$$s^2=[n_1n_3/(N^2-N)][(N^3-N)]/(12-S_t)]$$

In the example this gives a result of $s=21.27$ ($s^2=452.586$), and so the corrected value of z is 0.940 (still not significant).

The effect of making the appropriate correction for ties will always be in the direction of increased significance, although usually only slightly so, so there is a small increase in power.

that no adjustment for multiplicity should be made to take account of the number of different weeks involved.

When it comes to taking account of the use of several different dose levels, however, some allowance for multiplicity is probably appropriate. One way to do this is to carry out a between group comparison which incorporates all groups in the same analysis. A suitable non-parametric procedure is the Kruskal–Wallis test, which gives a single statistic (usually referred to as 'H') which tests for the presence of differences somewhere among the groups. One way to use this test to guard against multiplicity is to carry out the Kruskal–Wallis test first, and only proceed to pair-wise comparisons using the Mann–Whitney U-test if the Kruskal–Wallis H statistic is significant. Alternatively several methods are available for making pair-wise comparisons within the Kruskal–Wallis procedure (see e.g. Hollander & Wolfe, 1973, Chapter 6 Section 3). Of the methods discussed by Hollander & Wolfe that due to Dunn (1964) seems the simplest to apply, and has the advantage that it copes easily with unequally sized groups.

The procedure described in the preceding paragraph, however, can be criticised because it does not take into account the ordering of the doses, whereas biologically an effect in the low dose group can have different implications from an effect in the high dose group. Also, the interpretation of an effect in the low dose group will depend on whether or not an effect is also present in the high dose group.

One method of taking account of the ordering of doses is to use a test specifically designed to detect situations in which the response increases (or decreases) with increasing dose level. Such tests are sometimes referred to as *trend tests*. A suitable non-parametric trend test is the Jonckheere procedure, illustrated in Box 8.5.

The Jonckheere procedure gives a result which simply tests whether or not there is evidence for a trend; it does not make comparisons between individual doses and the control. An alternative procedure, which first tests for an effect at the high dose level and then proceeds to look at the next level down only if an effect is found at the top level, is given by Shirley (1977). In some situations (e.g. when the top dose decreases fertility) it is possible for the maximum effect to occur at the next to top dose instead of the top dose itself. In such cases, care should be taken with the trend tests, perhaps excluding the top dose from the analysis.

8.3.3 Parametric approaches based on the normal distribution

The non-parametric procedures described above are useful, but they are limited because they cannot easily be extended to take account of more complex protocols or treatment comparisons. Also, if the study has been designed using blocking, the potential gain in precision available cannot be realised. The technique of analysis of variance does not suffer from these restrictions, and it can be applied in such a way as to take proper account of the within and between male variability in a study in which more than one female per male per week is used (by using a *hierarchical* analysis). In this chapter, however, this complication is ignored and the technique is illustrated in Box 8.6 by application to data averaged for each male (this is quite valid for testing between treatment effects, but not for between week effects).

Analysis of variance techniques, however, cannot validly be applied to data which are markedly non-normal in their distribution. Parameters such as the average number of implants per female are usually close enough to a normal distribution for the analysis of variance to be used on the data as they stand. For other parameters, such as the proportion of early deaths, some skewness is likely to be present and it is usual to make a transformation of the data before doing the analysis of variance. A wide variety of possible transformations has been suggested in the literature, and some of these are easier to use in practice than others.

For parameters which are counts rather than proportions, such as the number of pregnancies per male (*PR*), a square root transformation is appropriate. A simple square root should be quite sufficient, but statistical purists prefer to take the square root of ($PR + 0.375$) (Anscome, 1948).

Box 8.5. *The Jonckheere test*

The first step is to calculate Mann–Whitney U statistics for all possible pairs of groups. Using the data from Table 8.1 as an example, there are four groups, so there are six possible pairs of groups and the required U values are:

$U_{12}=125 \quad U_{13}=125 \quad U_{14}=143 \quad U_{23}=111 \quad U_{24}=142 \quad U_{34}=121.5$

(Be careful in each case to treat the lowest numbered group of the pair as equivalent to the control group in the example in Box 8.1 e.g. when calculating the U value for the low and high dose groups calculate U_{24} and not U_{42}).

The Jonckheere statistic, J, is the sum of all the U values, which is equal to 767.5 in the example.

To assess the significance of J, calculate

$z=(J-m)/s$

where

$m=[N^2-\sum\{n_j^2\}]/4$

$s^2=[N^2(2N+3)-\sum\{n_j^2(2n_j+3)\}]/72$

$\sum$ denotes summation over all groups

n_j denotes the number of males for which data are available in group j.

$N=\sum n_j$

In the example, $n_1=n_2=15$, $n_3=n_4=14$, $N=58$, $m=630.5$, $s=72.01$, and hence $z=1.90$. This is greater than 1.64, the value required for significance at the 5% level for a one-tailed test (using tables of the standard normal deviate, z), and so a significant trend is present in these data.

Another modification sometimes used is the Freeman–Tukey Poisson transformation (Freeman & Tukey, 1950; Mosteller & Youtz, 1961),

$$FTP=\surd(PR)+\surd(PR+1)$$

but this suffers from the difficulty that it is not easy to back-transform the treatment means when the analysis is completed.

For parameters which are proportions a bewildering array of possibilities exists. These are all based on the *angular* transformation, also called the *arc-sine* transformation,

$$AS=\sin^{-1}\surd(ED/TI)$$

$ED=$ Early deaths

$TI=$ Total implants

This transformation, however, and also a modification of it which involves using a different equation when $ED=0$ or $ED=TI$, has been shown not to

Box 8.6. *The analysis of variance*

The detailed calculations for the analysis of variance are not given here, because most scientists have access to a computer program which will carry out this analysis. Here we give the results of a simple one-way analysis of variance carried out on the data given in Table 8.1. The data were averaged by dividing the total number of early deaths by the total number of implants for each male, with the resulting proportions then being transformed using the *CH* transformation with $c = 0.025$ as suggested in the text. In this example, as many decimal places as possible were retained for the intermediate results (as is good computational practice), except that the values for *CH* were rounded to two decimal places. This was done for the convenience of anyone trying to reproduce these results with limited computing facilities; if all the calculations can be done in a single package it would be counter-productive to deliberately round the *CH* values. The difference in the results caused by the rounding of the *CH* values does not become apparent until the fourth significant digit.

The first male in the intermediate dose group in the data in Table 8.1 has a total of one early death out of 26 implants, so $1/26 = 0.03846$ is the untransformed mean ratio. The transformation gives

$$CH = \sin^{-1}\sqrt{(0.025 + 0.95 \times 0.03846)} = \sin^{-1}(0.24807) = 14.36^{\circ}$$

This transformation has been carried out on all the data in Table 8.1; all the occasions where no early deaths were recorded result in a transformed value of 9.10°. The two animals recording no pregnancies were excluded.

Analysis of variance table

Source	Degrees of freedom	Sum of squares	Mean square	*F* ratio	*P*-value
Treatment	3	116.67	38.89	1.61	0.20
Residual	54	1307.46	24.21		
Total	57	1424.13			

Note:
If the study was designed in blocks, blocks are included as a factor in the analysis, making it a two-way analysis of variance.

Table of (transformed) means

Group	Control	Low	Intermediate	High
Mean	12.65	13.17	14.18	16.37
SE	1.27	1.27	1.32	1.32
n	15	15	14	14

Notes:
SE: standard error of mean.
n: number of observations.

Note that the standard errors given above use the pooled estimate of error derived from the analysis of variance, and are not the same as the

Box 8.6. (*cont.*)

standard errors derived using only the data from the individual treatment groups.

Comparisons between means

Most computer packages give the standard errors of the group means, given above. The most useful statistic for testing the difference between group means, however, is the standard error of a difference between means, which is not usually given. This is calculated, for a particular pair of means with n_i and n_j observations respectively, as

$$SED = \{RMS(1/n_i + 1/n_j)\}$$

where *RMS* is the residual mean square (or error mean square) from the analysis of variance (24.23 in the example above). Thus for the comparison between the control and intermediate group means in the example,

$$SED = \{24.23(1/15 + 1/14)\} = 1.829$$

The same *SED* applies to the comparison of the means for the control and high dose groups, but for the control and low dose groups the *SED* is slightly different (1.797) because both groups have 15 observations.
To test the difference between means, simply divide the difference by its *SED*; this gives a value which can be looked up in tables of the Student t distribution (or, better, put into a computer program to obtain a P-value). Thus in the example, the difference between the control and intermediate dose group means, on the transformed scale, is 1.53. Divided by the appropriate *SED* (1.829), this gives the t value of 0.837. The appropriate number of degrees of freedom for finding the significance level, or P-value, is the number of degrees of freedom for the residual in the analysis of variance (54 in the example). This is not significant at the 5% level, and the exact P-value, obtained from a computer program, is 0.203 (one-tailed).
The full set of results is best presented as back-transformed means with the P-values for comparisons with the control group. The back-transformed means can be calculated by applying the transformation in reverse:

$$[(\sin^2 y) - c]/[1 - 2c]$$

where y is the transformed mean.

Table of (back-transformed) means

Group:	Control	Low	Intermediate	High
Mean	0.024	0.028	0.037	0.057
P-value	—	0.39	0.20	0.023

Note:
P-value: One-tailed P-value for comparison with the control group.

Interpretation

In the example, the high dose group shows a statistically significant increase in the proportion of early deaths in the top dose group compared with the controls, but not at the low and intermediate dose groups. Such a

Box 8.6. (*cont.*)

result should be taken as an indication of a treatment effect, even in the absence of a significant result for the F ratio in the analysis of variance. If the only significant comparison with the control occurs in the low or intermediate dose groups, however, a higher level of significance (lower P-value) would be required before claiming a real treatment effect. This is an area where close collaboration between the biologist and statistician is necessary in order to make the most reasonable interpretation.

achieve the desired objective of making all treatment groups have similar variances when some of the proportions are close to zero or 100% (Chanter, 1975).

Two other versions of this transformation, that suggested by Anscombe (1948) and the Freeman–Tukey Binomial (Freeman & Tukey, 1950; Mosteller and Youtz, 1961), both suffer from the problem that the result depends on the denominator; thus 1 out of 10 transforms to a different value from 2 out of 20. In data where the denominator can vary, as is the case with the dominant lethal assay, this property seems undesirable. The Freeman–Tukey Binomial also suffers from the same back-transformation problem as its Poisson counterpart.

One transformation which does not suffer from these problems is the following (Chanter, 1975):

$$CH = \sin^{-1}\sqrt{\{c + (1 - 2c)ED/TI\}}$$

where c is a small positive quantity. Its exact value is not critical, and can be taken to be 0.025 for typical applications in this field. It should be noted, though, that in situations in which the denominator is variable, transformations of this kind do not remove heterogeneity of variance arising as a result of this; this problem could be overcome by using a weighted analysis of variance (Chanter, 1975; Haseman & Kupper, 1979).

Having selected and used an appropriate transformation, the analysis of variance is then carried out in the usual way. If a suitable computer program is available much of the numerical work can be avoided (the more sophisticated packages will do the transformation as well). As an illustration a typical analysis of variance table is shown for a one-way analysis of variance of data for one week's mating, using the CH transformation, in Box 8.6. Many computer packages are capable of doing a two-way analysis of variance which includes data from all weeks in the same analysis. However, unless this is set up in a special way, so that the variance appropriate for making comparisons between weeks is not confused with that appropriate for making comparisons between groups, the results could be seriously misleading. It is recommended, therefore, that such two-way analyses of variance should be used only under the guidance of a qualified statistician.

Following the analysis of variance, comparisons between the treatment means can be made, as with the non-parametric analyses, according to a variety of philosophies. The simplest approach is to test the difference between each pair of treatment means using Student's t distribution. This procedure is illustrated in Box 8.6, and is loosely analogous to the repeated Mann–Whitney U-tests described earlier in terms of its false positive error rate properties. When a transformation has been used, the comparisons between means should be carried out on the transformed scale, although back-transformed means, on the original scale, can be calculated to assist with the interpretation (see Box 8.6).

There are many procedures, collectively known as 'multiple comparison' procedures, which have been designed to protect against multiplicity (i.e. carrying out too many tests and hence getting too many false positive results). However, the vast majority of these have been designed for the situation where there is no treatment structure. In most dominant lethal assays, there is a treatment structure (usually increasing dose levels, including zero), and so most of these multiple comparison procedures are inappropriate. One procedure which has been specifically designed for the control and several increasing dose levels situation, however, is that due to Williams (1971, 1972). The equivalent non-parametric technique (Shirley, 1977) mentioned earlier, is based on this approach. It is recommended (except in the situation described in the next paragraph) that Williams' test be used, under the guidance of a qualified statistician, if there is any unease about using the series of Student t comparisons illustrated here.

Both Williams' and Shirley's tests are based on the assumption that the treatment effect, if present, will not in reality reverse its direction as the dose is increased (individual treatment means may be 'out of order' relative to the dose levels, but this is assumed by the test procedure to be the result of random variation). Thus if there is any biological reason to expect a reversal of the treatment effect, or if the data very strongly suggest this, these approaches should not be used.

8.3.4 'Cluster' analysis

The use of a transformation prior to analysis of variance has the twin objectives of stabilising variance (i.e. making the variance of all groups about the same) and improving the normality of the distribution within each group. For data with a substantial proportion of zeros (as is usually the case with ED/TI), transformations are not particularly effective with the second of these objectives, and an alternative method of achieving normality has been suggested by Whorton (1981).

Whorton's approach consists of grouping the males within each treatment group, at random, into a series of 'clusters' and calculating a mean value for each cluster. These clusters are then regarded as the basic experimental units. Whorton recommends clusters of size 6.

This approach has some serious difficulties associated with it. Perhaps the most serious of these is that the analysis is not reproducible; another analyst, working with the same individual animal data, would get different answers because his randomisation would be different and would lead to different clusters being defined. In our view, this is a more serious problem than that posed by the presence of an appreciable proportion of zeros in the analysis of variance, and we do not therefore recommend this approach.

8.3.5 Alternative approaches

A number of alternative approaches, including parametric methods based on the binomial distribution, are reviewed by Haseman & Kupper (1979). In particular, methods based on the binomial distribution (more precisely, beta-binomial), seem particularly appropriate because they attempt to model the probabalistic structure of the assay, do not lose any information by ranking (as the non-parametric methods do) and do not require a transformation (as methods based on the normal distribution do). Smith & James (1984) compare the fit of several different theoretical distributions to some real data. The assistance of a qualified statistician should be sought if these methods are to be used.

8.4 CONCLUSIONS AND RECOMMENDATIONS

To a large extent the choice of which method of statistical analysis to use is a matter of personal taste. The non-parametric methods are preferred by some workers because of their lack of dependence on assumptions about the distributional form of the data. The normal distribution methods are preferred by others because of their flexibility. The beta-binomial methods are preferred by yet others because the assumptions involved seem the most reasonable.

Fortunately, for the majority of data sets, all methods lead to the same (or very similar) conclusions, and so the choice of method is not crucial. It is only in equivocal cases, where the experimenter is uncertain as to whether a treatment effect is indicated or not, that different methods might lead to formally different interpretations. In such cases, however, the best interpretation is that the data are equivocal. If this is considered unsatisfactory, there is always the option of a repeat study, although it would be unethical to repeat the study without first having made a thorough examination of the data from the first study, involving more than one method of analysis.

We do not, therefore, recommend any particular method as superior to any other. Rather, the advice is that the experimenter should use whichever method he or she feels most comfortable with. We do, however, consider it important to stress the following points:

(a) Make sure that the design of the assay is sound and efficient. Have your design checked by a qualified statistician, giving him or her as much information as you can about sources of variation in your experimental material.

(b) Don't just feed your data into a computer program for the analysis; look at the data carefully, plot them in any way you think might be informative, and think about such things as whether any individual data points might be having an unduly large effect on the results of the statistical analysis.

(c) If you are unsure about any aspect of the analysis of your data, consult a statistician.

8.5 REFERENCES

Anderson, D., Bateman, A. & McGregor, D. (1983). Dominant lethal mutation assays. In *UKEMS Sub-committee on Guidelines for Mutagenicity Testing*, Report, Part I, Basic Test Battery, Ed. B.J. Dean, United Kingdom Environmental Mutagen Society, Swansea pp. 143–64.

Anscombe, F.J. (1948). The transformation of Poisson, binomial and negative-binomial data. *Biometrika*, **35**, 246–54.

Bateman, A.J. (1984). The dominant lethal assay in the male mouse. In *Handbook of Mutagenicity Test Procedures*, (Eds. B.J. Kilbey, *et al.*) Elsevier, N. Holland.

Chanter, D.O. (1975). Modifications of the angular transformations. *Applied Statistics*, **24**, 354–9.

Dean, B.J. and Johnstone, A. (1977). Dominant lethal assays in male mice: evaluation of experimental design, statistical methods and the sensitivity of Charles River (CD1) mice. *Mutation Research*, **42**, 269–78.

Dunn, O.J. (1964). Multiple comparisons using rank sums. *Technometrics*, **6**, 241–52.

Freeman, M.F. and Tukey, J.W. (1950). Transformations related to the angular and the square root. *Annals of Mathematical Statistics*, **21**, 607–11.

Haseman, J.K. and Soares, E.R. (1976). The distribution of fetal death in control mice and its implications on statistical tests for dominant lethal effects. *Mutation Research*, **41**, 277–88.

Haseman, J.K. and Kupper, L.L. (1979). Analysis of dichotomous response data from certain toxicological experiments. *Biometrics*, **35**, 281–93.

Healey, G.F. (1983). Statistical contributions to experimental design. In *Animals and Alternatives in Toxicity Testing* (Ed. M. Balls, R.J. Riddell & A.N. Worden), Academic Press, London, p. 167–88.

Hollander, M. and Wolfe, D.A. (1973). *Nonparametric Statistical Methods.* Wiley, London.

Lovell, D.P., Anderson, D. and Jenkinson, P.C. (1987). The use of a battery of strains of mice in a factorial design to study the induction of dominant lethal mutations. *Mutation Research,* **187**, 37–44.

Mosteller, F. and Youtz, C. (1961). Tables of the Freeman-Tukey transformations for the binomial and Poisson distributions. *Biometrika,* **48**, 433–40.

Shirley, E. (1977). A non-parametric equivalent of Williams' test for contrasting increasing dose levels of a treatment. *Biometrics,* **33**, 386–9.

Siegel, S. (1956). *Nonparametric Statistics for the Behavioural Sciences.* McGraw-Hill.

Smith, D.M. and James, D.A. (1984). A comparison of alternative distributions of postimplantation death in the dominant lethal assay. *Mutation Research,* **128**, 195–206.

Whorton, E.B. (1981). Parametric statistical methods and sample size considerations for dominant lethal experiments: the use of clustering to achieve approximate normality. *Teratogenesis, Carcinogenesis and Mutagenesis,* **1**, 353–60.

Williams, D.A. (1971). A test for differences between treatment means when several dose levels are compared with a zero dose control. *Biometrics,* **27**, 103–17.

Williams, D.A. (1972). The comparison of several dose levels with a zero dose control. *Biometrics.* **28**. 519–31.

9

Statistical methods for the design and analysis of mutation experiments with the fruit fly *Drosophila melanogaster*

R.D COMBES (Group Leader)
J. BOOTMAN M.G. FORD
J. HEPWORTH D.W. SALT

9.1 INTRODUCTION AND SCOPE OF CHAPTER

The use of Drosophila for genotoxicity testing has been described in numerous reviews (see, for example, Wurgler *et al.*, 1984; Mitchell & Combes, 1984) and detailed recommendations for the sex-linked recessive lethal test have been documented in an earlier volume of these reports (Bootman & Kilbey, 1983). Drosophila can be used to detect recessive lethal mutations which comprise point mutations and deletions. Different test systems allow the detection of both point mutations and deletions in somatic and germ cells and the scoring of other end points such as non-disjunction, chromosome loss and translocation. In this chapter, we are only concerned with germ cell mutation as detected by the sex-linked recessive lethal assay: this is by far the most commonly undertaken test. Our main objective is to suggest appropriate statistical methods for discriminating between mutagenic and non-mutagenic compounds and to use statistics to recommend appropriate methodological designs to obtain optimum sample sizes and to reduce the time and cost of experiments when using the sex-linked recessive lethal assay. However, the principles of experimental design and statistical analysis which are discussed may apply to the other assays mentioned above.

9.2 THE SEX-LINKED RECESSIVE LETHAL ASSAY

9.2.1 Introduction

In order to design mutation experiments and to apply the most appropriate statistical analyses for the data, it is necessary to be conversant with the theoretical background to the sex-linked recessive lethal assay.

9.2.2 Theoretical considerations

Wild type male flies (Fig. 9.1) which have been exposed to the test agent are mated individually to Muller 5 virgin females whose *X*-chromosomes bear the white-apricot (light orange) eye recessive allele (w^a) and the semi-dominant bar eye mutation (*B*). In this way an *X*-chromosome from a male which carries a lethal mutation is transmitted to the F_1 progeny. Each individual female inheriting this chromosome represents one *X*-chromosome treated in the paternal male and is mated with a Basc male to give the F_2 generation. The F_1 progeny from each treated male are kept together by labelling vials so as to identify all mutations arising from a common male. Individuals carrying a recessive lethal mutation die when the mutation is not masked by a dominant allele on the corresponding homologous chromosome. Thus, although F_2 females with such a mutation on one of their two *X*-chromosomes survive, males with a single *X*-chromosome bearing a recessive lethal mutation with no corresponding allele on the *Y*-chromosome do not arise. Recessive lethal mutations cannot have occurred if vials contain F_2 flies which include wild type males; these males must carry the *X*-chromosome from the original paternal wild type male. The number of lethal mutations is equal to the number of vials which lack wild type male flies since these are absent as a result of inheritance of a recessive lethal mutation (see Fig. 9.1). This method of distinguishing lethals can be adopted only if exposed males are mated separately to females to give the F_1. If this is not the case, it is likely that, even though one of the males carried a lethal mutation, others in the same vial will not. As a consequence, wild type males will be present in F_2 vials despite such vials being derived from a male carrying a lethal mutation.

9.2.3 Experimental design and presentation of data

Data from the sex-linked recessive lethal assay can be expressed as the relative frequency of transmitted mutations calculated as the number of *X*-chromosomes with a lethal divided by the number of *X*-chromosomes exposed (number of vials containing flies) (see Tables 9.1 and 9.6).

A detailed protocol for the test is presented in Mitchell & Combes (1984). Important points concerning the design and implementation of the

Fig. 9.1. Genetic basis of the sex-linked recessive lethal assay in Drosphila.
+ denotes the wild type (dominant) allele of the eye gene; w^a denotes the mutant (recessive) allele of the eye gene resulting in white apricot eye colour; *B* denotes the mutant allele of the bar eye gene affecting eye shape. This allele is incompletely recessive such that in *B*/+ heterozygotes the eye is kidney shaped, and bar shaped in *B*/*B* double mutant homozygotes or when in the hemizygous condition in the male (*B*−).

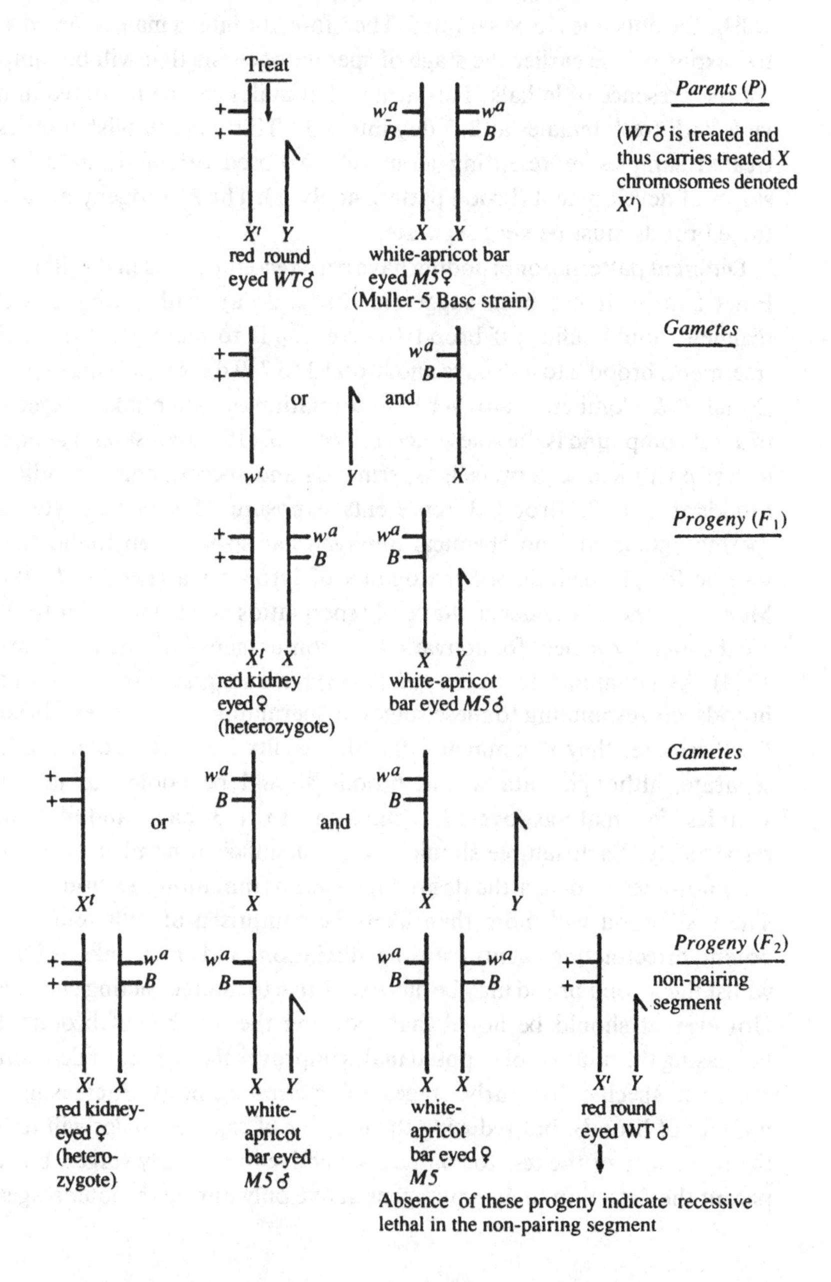

protocol are detailed in Section 9.4.1.1. For purposes of experimental design and statistical analysis, two further considerations must be taken into account. The first relates to differential sensitivity of the germ cells. Spermatogenesis lasts for nine days and certain stages of this process vary in their sensitivity to mutagens. If males are mated once, immediately after test exposure, cells that were mature spermatozoa at the time of treatment will be the only ones to be sampled. Therefore, the later a male is mated after test exposure the earlier the stage of spermatogenesis that will be sampled for the presence of lethals. This means that males can be re-mated in new vials with fresh females at 2–3 day intervals. This will establish a series of treated samples representing germ cells exposed originally at different stages of development (brood pattern analysis). The F_2 progeny of each of these broods must be kept separate.

Different patterns for brooding have been recommended in the literature. For example, it has been suggested that a 3-day, rather than a 4-day, regimen should suffice, if brood 1 corresponds to mating 1-3 days after treatment, brood 2 to 4–6 days and brood 3 to 7–9 days following exposure (Mitchell & Combes, 1984). Where information on potential mutagenicity of a test compound is the sole concern, Lee *et al.* (1983) consider it adequate to test post-meiotic germ cells (spermatids and sperm), corresponding to broods 1 and 2. Brood 3 represents exposure of spermatocytes and spermatogonia and no chemical mutagen has so far been found that is specific for pre-meiotic spermatogonia of Drosophila (Lee *et al.*, 1983). Moreover, there is evidence that early spermatids and stages prior to these are the most proficient for activation of promutagens (Mitchell & Combes, 1984). As a consequence, Lee *et al.* (1983) have suggested that at least two broods, corresponding to these stages of spermatogenesis, are established. Furthermore, they recommend that the results for each brood are kept separate, although data within broods should be pooled to give two samples for matings overall, equivalent to 1–3 days and 4–7 days respectively. Each sample should involve a sufficient number of exposed chromosomes to detect the desired increase in mutation (Section 9.4.1.1). The first brood will more than likely be comprised of cells mutated by potent, direct-acting agents (causing alkylations and cross-links in DNA), whilst the second brood may be more sensitive to indirect-acting chemicals. However, it should be noted that reducing the number of broods, but increasing the number of exposed males improves the chance of detecting a mutagen specific for early stages of spermatogenesis. Increasing the number of broods, but reducing the number of exposed males will reduce the sensitivity of the test to compounds active in the early stages, but will permit the detection of any mutagens active only during the later stages of sperm development. The choice of brood number will therefore depend on the objectives of the experiment. This working group discussed these options and advises the use of three broods always for routine testing of unknown chemicals, where the specificity of such compounds to germ cell stage cannot be predicted. It may be necessary to base the estimation of mutation rate on pooled data from the three broods in order to achieve a statistically significant result.

A second related problem is due to clustering of mutations. If a spontaneous or induced lethal mutation arises in an early, pre-meiotic stage of sperm development (e.g. spermatocytes), several mutant chromosomes of common origin will be inherited giving rise to a clone or cluster of progeny bearing identical recessive lethals following the induction of only one original mutation. It is essential that these clusters can be distinguished from an individual lethal mutation. The only way that this can be achieved is by labelling vials in such a way that it is possible to relate each treated *X*-chromosome in the progeny to a particular exposed male. Each original male should give rise to no more than one F_2 vial which indicates a lethal mutation because it does not contain a wild type male. When clusters do arise they have to be accounted for in the statistical analysis and the way in which they have been dealt with must be recorded (see Section 9.2.4).

Many investigators seem content to score mutant clusters as single mutational events so as to avoid the additional statistics which have been recommended for their analysis (Muller, 1952; Engels, 1979). However, this ignores the possibility that more than one lethal could arise from one male due to two independent mutational events occurring in different germ cells within the same fly. This working group endorses the recommendation of Bootman & Kilbey (1983) that, for each cluster, every mutation should be treated as an individual mutant in the initial statistical analysis. If the data are non-significant, then no further analysis is required. However, if positive mutagenicity is indicated, then the data should be re-analysed whilst treating each cluster as a single mutational event. If, after this, the results retain significance, the data should be considered indicative of mutagenicity since such an analysis will give an under-estimate of mutation. However, if the figures become non-significant, the mutation test should be repeated to determine whether the compound has a propensity for inducing clusters.

9.2.4 Published criteria adopted for positive and negative effects

Protocols commonly used for the sex-linked recessive lethal assay are not particularly sensitive from a statistical point of view, since approximately a seven-fold increase in the frequency of spontaneous mutation is required for detection at the 5% significance level ($P=0.05$). Although sensitivity can be improved by changes in methodology, the main problem is coping with the large numbers of flies that are needed to achieve this objective.

An Environmental Protection Agency (EPA) working group (Lee *et al.*, 1983) published criteria for mutagenic and non-mutagenic effects for the sex-linked recessive lethal assay. Experiments which included concurrent negative controls and tests where these were omitted were discussed. In the former case, a compound was considered mutagenic if the difference in frequencies between treated and control was statistically significant at $P=0.05$ according to Kastenbaum-Bowman tables (Section 9.4.3), irrespective of whether or not results for individual broods had been pooled. Data obtained with flies from individual broods do not often achieve significance because insufficient numbers of progeny are scored. In this case no definite conclusions concerning sperm cell stage specificity of the compound can be derived from the data.

In the second situation, as spontaneous frequencies for the normal tester strains lie between 0.1 and 0.3%, Lee *et al.* (1983) considered a frequency of 0.5% to be indicative of mutagenicity. The same authors classified a compound as non-mutagenic if treatment resulted in an increase in mutation over the control of less than 0.2% providing that adequate sample sizes were used to permit a 0.2% increase to have been statistically significant. In contrast with the suggestions of Lee *et al.* (1983), this working group strongly recommends that only data from experiments where a negative control is conducted concurrently should be used for analysis.

Crow & Temin (1964) calculated the overall spontaneous mutation rate as 0.26% but noted some interlaboratory variation. Lee *et al.* (1983) suggest that, in order to conclude that a compound is non-mutagenic, no statistically significant increases in mutations should be detected in the treatment group compared with the control in any of the broods which cover exposure of all the different germ cell stages. This is to ensure that inactivity cannot be due to exposure of sperm cells at insensitive stages of development. They emphasise that these criteria for mutagenicity differ from those normally used which stipulate an increase of 0.2% in the treated group over the control. These new criteria allow for variations in spontaneous mutation rates to be taken into account. However, if concurrent negative controls are included, this will be unnecessary.

Sample sizes required to detect particular increases in the mutation frequency with varying negative control rates are presented in Wurgler *et al.* (1977), although these sizes are not recommended by Margolin *et al.* (1983).

Lee *et al.* (1983) discuss the sample sizes in each of their recommended broods 1 and 2 (above) when the numbers from each brood are pooled.

9.2.5 Published statistical methods used for Drosophila

The most widely used statistical method for analysis of Drosophila mutagenicity data is that devised by Kastenbaum & Bowman (1970), based on the binomial distribution. The application of Kastenbaum–Bowman tables has been described in detail by Wurgler *et al.* (1977, 1984). This analysis is equivalent to the more detailed, but non-parametric, Fisher's exact test (Berchtold, 1975). It is important to remember that Kastenbaum–Bowman tables should only be applied if the total number of mutations recovered in both treated and negative control populations remains below 100 (Lee *et al.*, 1983). With higher numbers of mutations, a simple chi-squared test is considered to be adequate. Certainly the latter test should not be used when the number of mutations is less than five. Recommendations for the use of statistical tests under particular circumstances are presented in this chapter (Section 9.4.4).

All statistical methods make at least two basic assumptions. Firstly, that observations are independent, and secondly that the probabilities of either outcome (mutagenicity or non-mutagenicity) are the same for every observation. Occasionally these criteria may be invalid. Test organisms can vary in their susceptibility to compounds so that mutagenic events occur with differing probabilities. Furthermore, mutants of clonal origin (clusters) do not arise independently. It can be difficult to ascertain the extent of differential sensitivity, especially with low numbers of mutants and other, non-parametric methods for analysis have been advocated (Mitchell & Combes, 1984). As mentioned earlier, clusters are traced back to each parental exposed male and can be eliminated from the data set using methods described by Engels (1979).

9.3 GENERAL CONSIDERATIONS FOR STATISTICAL ANALYSIS

9.3.1 Introduction

Mutagenicity data may be tabulated (see Section 9.4.2 and Tables 9.1 and 9.2) to give the numbers of mutants recovered in a control population of a certain size and numbers of mutants in a treatment population of a given size. Population sizes relate to the number of control or treated male flies used which is the same as the number of unexposed and exposed *X*-chromosomes (Fig. 9.1). Therefore the data are in the form of a 2×2 contingency table, one category comprising the treatment versus

Table 9.1. *An example of presentation of mutation data in the format shown in Table 9.2.*

	Population numbers		
	Control	Treated	Totals
Mutant *X*-chromosomes	4	11	15
Non-mutant *X*-chromosomes	1716	1709	3425
Totals	1720	1720	3440

Note:
An example of the analysis of this data set using the normal test is described in Section 9.4.3.5.

spontaneous (negative control) mutation rate and the second category being positive mutagenicity versus negative mutation rates. Analysis of such data will not be straightforward if the test chemical induces mutagenic responses at a low frequency, for example below 0.01 (1 mutant in 100 exposed male flies, or *X*-chromosomes). Investigators should, therefore, always be prepared to check that assumptions underlying standard statistical analyses are valid, and if not they should be prepared to use alternative methods (see Section 9.4.4).

9.3.2 Hypothesis testing and error type

The reader will find a discussion underlying the philosophy of hypothesis testing in Chapter 1 in this volume. However, the recommendations of this working group require a detailed explanation of the power of statistical tests. The following account provides the essential background theory for deciding which of the available statistical tests is most appropriate for analysing data from experiments with Drosophila in a given situation.

9.3.2.1 *Null and alternative hypotheses*

One of the objectives of statistical analysis of mutation data is to determine whether the mutation rate (p_t) in the treated group, comprising

Table 9.2. *Tabular form of results for a mutagenicity experiment*

	Experimental groups		
Observation	Control	Treatment	Marginal totals
Yes	X_c	X_t	M
No	$N_c - X_c$	$N_t - X_t$	$N-M$
Marginal totals	N_c	N_t	N

N_t experimental units is larger than the mutation rate (p_c) in a control group comprising N_c experimental units. The classification of a candidate compound as mutagenic or non-mutagenic is usually based on some form of statistical hypothesis test in which the null hypothesis (H_0) is stated as H_0:$p_t = p_c$ (or equivalently $p_t - p_c = 0$) and the alternative hypothesis (H_1) is stated as H_1:$p_t > p_c$ (or equivalently $p_t - p_c = \delta > 0$), where δ is the difference between the spontaneous mutation rate and the rate due to exposure. The null hypothesis assumes that treatment with the candidate compound results in the same mutation rate as the control and therefore that the compound can be classified as being non-mutagenic. The alternative hypothesis assumes that exposure to the chemical results in a higher mutation rate and that it can therefore be classified as being mutagenic. This form of hypothesis test is called a one-sided (or one-tailed) test and it is considered to be more appropriate in genetic toxicology testing where the experimenter is interested primarily in the directional mutagenic alternative $p_t > p_c$ rather than in the less likely possibility where $p_t < p_c$ (an anti-mutagenic effect).

Acceptance or rejection of the null hypothesis depends upon whether a quantity known as the *test statistic* lies within the critical region for a particular test (see Fig. 9.2 and Sokal & Rohlf, 1981). In the case of the one-sided test based on the normal approximation to the binomial distribution (see Section 9.4.3) where there are X_c and X_t observed mutations in the control and treated groups respectively, the test statistic (Z) can be expressed as:

$$Z = (p_t - p_c)/\sqrt{[p(1-p)(1/N_t + 1/N_c)]} \quad (1)$$

where p_t and p_c in equation (1) are the two observed proportions X_t/N_t and X_c/N_c respectively, and where p (the best estimate of the spontaneous mutation rate, assuming H_0 is true) is given by:

$$p = (X_t + X_c)/(N_t + N_c) \quad (2)$$

The alternative hypothesis that the mutation rate of the test compound is greater than the spontaneous rate is accepted if the test statistic Z falls in the critical region. In the case of the above one-sided test this corresponds to values of Z greater than a critical value, z_α, found from standard normal tables; where the lower case character (z) denotes particular values of the random variable in its general form (Z).

9.3.2.2 *Type I error*

α is the significance level of the test and is the probability of rejecting incorrectly the null hypothesis when in fact it is true. This is termed

Fig. 9.2. **Illustration of the relationship between α (probability of a type I error being committed) and β (probability of a type II error being committed) when testing the null hypothesis H_0: $P_t = P_c$ against the one-sided alternative hypothesis H_1: $P_t - P_c = \delta$. α and β are the blank and shaded areas under the H_0 and H_1 curves respectively. It can be seen that an increase in the area α will result in a reduction in the area β and vice versa.**

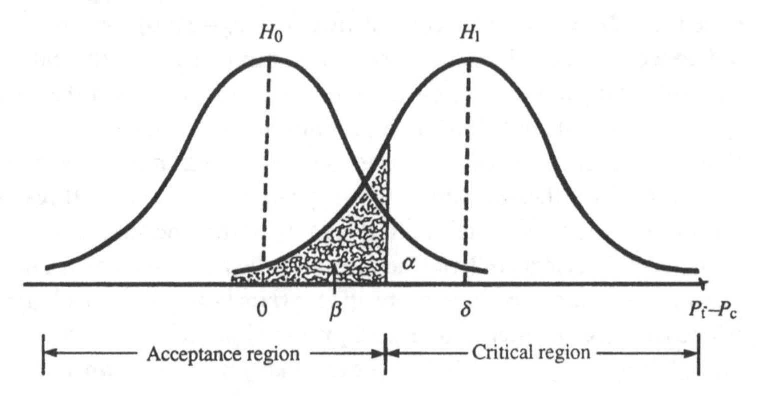

the Type I error and is equivalent to declaring a compound to be mutagenic when it is not (Fig. 9.2). This is illustrated in Table 9.3, where it can be seen that, if the compound being tested is non-mutagenic, that is H_0 is true and the decision to reject H_0 is incorrect, then a Type I error will have been committed. In mutation tests, values for α of 0.05 and 0.01 (5 in 100 and 1 in 100, respectively) typically are chosen, although others may be used. The smaller the risk of rejecting H_0 when it is true (i.e. committing a Type I error) the smaller the α value. Type I error can only be eliminated completely by making α zero. However, since this is equivalent to abiding by the commandment 'Thou shalt not reject H_0 whatever the evidence of the data', setting alpha at zero is not sensible.

9.3.2.3 *Type II error*

Unfortunately Type I error should not be the only concern when assessing mutagenicity test data. A second error, called Type II, is also present. This is the error committed when H_0 is accepted when it is false, that is an experiment fails to detect a compound that is in fact mutagenic (Fig. 9.2, Table 9.3). The probability of committing a Type II error, denoted by β, is the probability of accepting that a test compound is non-mutagenic when it is in fact mutagenic. The probability of rejecting H_0 when it is false is given by $(1 - \beta)$ which is the *power* of the test. α and β are not independent probabilities, and for given values of p_c, p_t, N_c and N_t, reducing α increases β and vice versa (Fig. 9.2). Therefore, by reducing the risk of classifying a non-mutagenic compound as being mutagenic, the risk of denoting a mutagenic compound as non-mutagenic is increased automatically.

Table 9.3. *Null hypothesis testing.*

Alternatives	Accept H_0	Reject H_0
	Correct decision	Incorrect decision
H_0 TRUE (test compound is non-mutagenic)	Non-mutagenic compound correctly detected with probability $(1-\alpha)$	Type I error made, non-mutagenic compound wrongly classed as mutagenic with probability α
	Incorrect decision	Correct decision
H_0 FALSE (test compound is mutagenic)	Type II error made, mutagenic compound incorrectly accepted as being non-mutagenic with probability β	Mutagenic compound correctly detected with probability $(1-\beta)$

A plot of $(1-\beta)$ against $N_c p_c$ $(=\lambda_c)$, with α, P_t and N_t held constant is known as the power curve for the particular statistical test (Fig. 9.3). Unfortunately it is common practice to select arbitrarily the number of test units for both treatment groups (N_c and N_t) together with a randomly selected α value without being aware of the existence of a Type II error and its likely implications. High values of β, and hence low power for the

Fig. 9.3. Power curves for the normal test plotted against λ_c $(=N_c p_c)$ (with a nominal $\alpha = 0.05$) for values of ρ $(=p_t/p_c)$ set to 2,3,4 and 5 (from Margolin *et al.*, 1983).

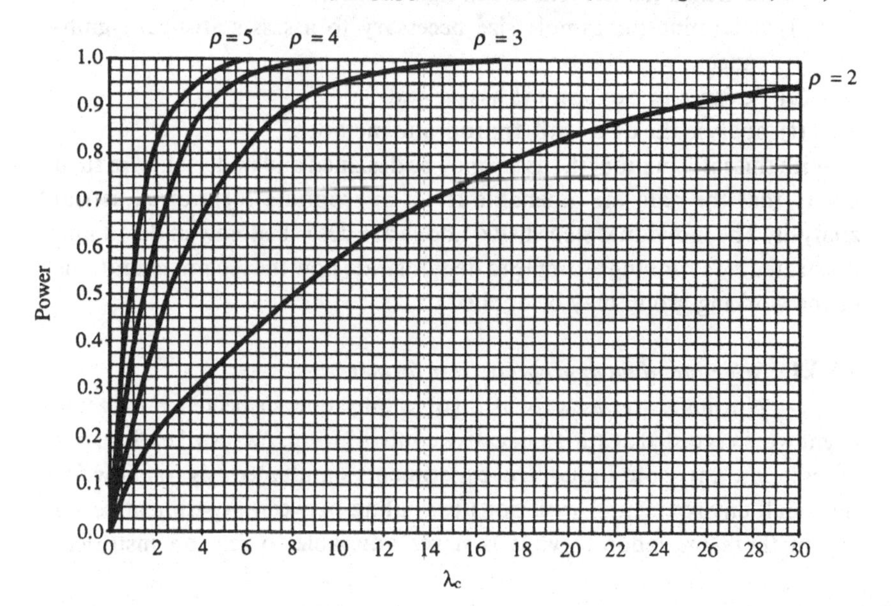

detection of a mutagen, imply that the reliability of the analysis is likely to be suspect (Sokal & Rohlf, 1981).

The significance level (α), quoted for a number of statistical tests currently in use for mutagenicity assays, such as Fisher's Exact test and the conditional binomial test (see Sections 9.4.3.1 and 9.4.3.2), is a nominal value. Because of the discrete nature of the test statistic in some binomial tests, sample size which determines the resolving power of the test can be critical. (For fixed α, increasing sample size increases resolving power). The true value of α (its effective value) can be considerably less than the critical value of α (its nominal value) specified at the design stage of the assay (Margolin *et al.*, 1983). Since a reduction in α will always result in a larger β, the power of the test $(1-\beta)$ can be extremely low, and the corresponding risk of accepting a mutagenic compound as being non-mutagenic will then be high. This is a disturbing feature of the design of mutagenicity tests where small sample sizes are used and where, consequently, some mutagens are likely to be classified as non-mutagenic.

9.3.3 Experimental design

When planning an experiment to detect a mutagenic compound the following regimen is recommended:

(a) select an appropriate number of treated and control groups (the test format);
(b) select the critical probability (the significance level);
(c) decide the smallest relative increase in mutagenicity at which it is desirable to show statistical significance;
(d) determine the sample size necessary to assess statistical significance;
(e) design and perform the assay, and
(f) analyse the data with appropriate methods.

With this information, the assay may be designed and undertaken in such a way that the data are amenable to an appropriate form of statistical analysis. The individual steps listed above are described now in detail and illustrated by a running example to demonstrate how the recommendations of the working group may be applied.

9.3.3.1 *Selection of an appropriate test format*

Normally the assay will involve only one suspected mutagenic agent which is tested against a negative control group. As stated earlier, it is necessary to carry out the control experiments simultaneously to allow for the small environmental variations prevailing between experiments even within the same laboratory. It is never advisable to rely on historical background mutation rates as a substitute for concurrent negative controls. The use of historical data should be restricted to a comparison of the reproducibility and repeatability of spontaneous mutation rates. The positive control is a known mutagen which can be helpful in indicating whether the assay is behaving as expected. Positive controls are particularly useful to indicate whether the tolerance of the test organisms varies either between laboratories or with time. Thus positive controls should be used in laboratories from time to time to check the sensitivity of the protocol. It is possible to test more than one suspected mutagenic agent against the same control group, and the procedure for analysis differs only slightly within the area of significance levels.

Running example

For routine testing, compare one treatment together with a negative control group and include a positive control group from time to time to act as an internal standard.

9.3.3.2 *Selection of critical probability values*

Testing for a difference between two groups involves determining the probability, α, that the observed differences between groups might have occurred by chance. When the probability of this happening by chance is low, the null hypothesis is rejected. The highest probability at which the null hypothesis is rejected is called the critical probability or significance level, and group differences are said to be significant at probability levels equal to, or less than, this critical probability value. Conventionally, a critical value of 1 in 20 ($\alpha=0.05$) is used, and less often 1 in 100 ($\alpha=0.01$). Other predetermined values can also be chosen and it is usually worth noting when differences very nearly attain significance. For instance, values between 0.05 and 0.10 are noteworthy when $\alpha=0.05$ and indicate that the test should be repeated and the decision of whether to accept a compound as mutagenic should then be based on the combined α probabilities (Sokal & Rohlf, 1981, p. 779). The working group considers this approach to be more satisfactory than that of Valencia *et al.* (1985) who chose $\alpha=0.01$ as the critical probability for a positive result, $\alpha=0.04$ for either questionable or positive (significant) results and $\alpha=0.06$ as equivocal (negative-insignificant). This categorisation is unsatisfactory because it gives the impression that the pre-selected significance level (α) can be modified subsequent to data collection.

Since the choice of critical probability level helps determine the size of the experiment (Section 9.3.3.4) it is important to select the value for α *before* the design of the experiment has been completed. Most statistical tables have values which are based on single comparisons (that is one treatment and one control) and two-tailed tests. There are two ways in which the criteria for these analyses differ in mutagenicity tests:

(a) The test should be one-tailed only (Section 9.3.2.1). Therefore probability values from two-tailed tables need to be doubled. For example, in considering a one-tailed test with $\alpha = 0.05$ it is necessary to look in the two-tailed tables under $\alpha = 0.10$.

(b) When several tests are being conducted and it is required that the overall significance level of the tests should be $\alpha = 0.05$ simultaneously, then corrections for the tests should be undertaken individually. When the number of tests is small such corrections can be straightforward. Two concurrent tests at $\alpha = 0.025$, for example, will give an overall test very nearly at level $\alpha = 0.05$. The Bonferroni corrections have been recommended for mutagenicity testing by Mitchell & Combes (1984). These corrections modify the significance level to account for the multiplicity of tests and thus reduce the risk of obtaining a significant effect by chance (Feller, 1971). It is only necessary to apply Bonferroni corrections, however, when a large number of treatments (e.g. > 10) is being assayed simultaneously.

Running example

Choose a critical, one-tailed probability value $\alpha = 0.05$.

9.3.3.3 ***Determination of the required level of mutagenicity***

It is necessary to decide on the level of mutagenicity that it is desired to detect. This is normally expressed in terms of the proportion of mutants arising after a fixed exposure relative to the proportion of mutants in a corresponding negative control group. For example, it may be considered appropriate to detect a treatment which causes a three-fold or four-fold increase in the background level of mutation. It is important to realise that, the higher the required level of distinction between treatment and control, i.e. the higher the ratio ($p_t/p_c = \rho$), the lower the detection sensitivity and the smaller the required sample size at the next stage of the experiment. Selection of this level (ρ), which is related to the power of the statistical analysis (Section 9.3.2.3), is very important and it is recommended that the sensitivity of mutagen detection, dictated by sample size, should be as great as constraints imposed by availability of laboratory resources and personnel permit. Historical control data should be used only at this stage, and not thereafter. Moreover, a strict protocol, based on these considerations, must be adhered to in all subsequent experiments. However, it should be borne in mind that if the necessary sample size still cannot be obtained it is better not to conduct the assay at all. It would be wrong to choose a sample size on the basis of a search for a three-fold increase and then find later a two-fold increase is required.

Running example

Suppose the background mutation rate (p_c)=0.0015 (i.e. 1.5 in 1000). Consider the detection of a three-fold increase in the mutation rate (then $p_t > 0.0045$ or 4.5 in 1000) and that the maximum sample size from resource limitations is 6000.

9.3.3.4 Determination of sample size

From both theoretical (Lehman, 1959) and practical viewpoints, there is a precedent for using equal sample sizes in the treatment and control groups, so that $N_c = N_t = N/2$. To determine the sample size in both groups the method of Margolin *et al.* (1983) should be used. Figure 9.3 represents plots of the power curve $(1-\beta)$ against $N_c p_c = (\lambda_c)$ for particular values of $\rho = p_t/p_c$, where the grid is imposed to allow easier interpolation. The use of this chart to obtain sample sizes is shown in the following illustration:

The spontaneous mutation rate (p_c) in a laboratory is 0.0015. An experiment is designed to detect a mutagen with a mutation rate four-fold that of the spontaneous rate ($\rho = p_t/p_c = 4$). The analysis will employ a significance level (α) of 0.05 and it is required that the mutagen be detected with a probability (power) of 0.90 (i.e. a four-fold increase in the mutation rate relative to that occurring spontaneously will just be detected at a significance level of 0.05 in nine out of every ten experiments). The minimum sample size for each group to satisfy the above given constraints is found from the chart in Fig. 9.3 by moving along the curve $\rho = 4$ until its ordinate = 0.90. The associated abscissa of 4.3 ($\lambda_c = N_c p_c$) divided by 0.0015 (p_c) indicates that a sample size of $N/2 = \lambda_c/p_c = 2867$ is required in each of the treated and control groups. Table 9.4 gives various examples calculated at a significance level of 0.05. For example, if the spontaneous mutation rate p_c is 0.0050 and it is necessary to detect a mutagen with a mutation rate three-fold greater than the control rate p_c ($\rho = 3$), using a significance level of 0.05 and a power of 0.80, then individual sample sizes of 1160 are required.

The chart in Fig. 9.3 may also be used to find the level of effect (ρ) of a mutagen expected at a given sample size. Suppose $p_c = 0.0015$ and the experimenter requires to detect $p_t = 0.0045$ using sample sizes

Table 9.4. *Individual group sizes required to achieve approximate power of 0.70, 0.80 and 0.90 for the nominal* $\alpha = 0.05$ *using the normal test with* $N_c = N_t$.

	Power					
	0.70		0.80		0.90	
	P_c		P_c		P_c	
ρ	0.0025	0.0050	0.0025	0.0050	0.0025	0.0050
2	5560	2780	7280	3640	9960	4980
3	1800	900	2320	1160	3200	1600
4	920	460	1280	640	1720	860
5	640	320	800	400	1120	560

$N/2 = N_c = N_t = 2000$ with a significance level (α) of 0.05. The required curve is $\rho = 0.0045/0.0015 = 3$. The abscissa value, $\lambda_c = N_c p_c = 3$ on this curve corresponds with an ordinate value of approximately 0.55. Thus there is only a 55% chance of detecting a mutagen that elevates the spontaneous mutation rate three-fold with sample sizes of 2000 and a significance level of 0.05.

For significance levels other than 0.05, the reader is referred to Margolin *et al.* (1983). The sample sizes quoted by these authors for power = 0.90 differ slightly from those presented in Table 9.4. These small differences arise because Margolin *et al.* (1983) computed their figures for the power curve with a mathematical formula, whereas the sample sizes in Table 9.4 are approximations obtained by interpolation from the chart in Fig. 9.3. In practice it is unlikely that exactly the same number of progeny in control and treatment groups will be analysed. Nevertheless, providing the ratio of actual treatment to control progeny is in the range 0.8 (e.g. $N_c = 2000$ and $N_t = 1600$ or $N_t/N_c = 1600/2000 = 0.8$) to 1.25 (e.g. $N_c = 2000$ and $N_t = 2500$ or $N_t/N_c = 2500/2000 = 1.25$) then no additional problems should arise.

Running example

Suppose $p_c = 0.0015$, $p_t = 0.0045$, $\alpha = 0.05$, $1 - \beta = 0.95$. For a 95% chance (i.e. power = 0.95) of detecting this three-fold increase in mutation (Fig. 9.3), $\lambda_c = 10.0$. This gives $N_c = N_t = \lambda_c/p_c = 10/0.0015 = 6667$. Because this value exceeds the maximum number that is economical it is necessary to re-design the experiment to enable detection of a four-fold increase in mutation rate i.e. $p_c = 0.0015$, $p_t = 0.0060$. Following the same procedure, $\lambda_c = 5.5$. and $N_c = N_t = 3667$ which is less than 6000 (Section 9.3.3).

9.4 APPLICATION OF STATISTICS TO THE SEX-LINKED RECESSIVE LETHAL ASSAY

9.4.1 Influence of assay protocol on choice of statistical test

In the sex-linked recessive lethal assay it is necessary to mate enough males with an adequate number of females to ensure that the size of the sample of sperm from an individual male is sufficient to increase the likelihood of detecting a statistically significant effect. If information on the sensitivity to potential mutagens at different stages of sperm development is required, it is important to obtain separate data for each of the broods. If the results for the three broods are pooled, this reduces the resolving power of the test. Therefore, the required sample size above should be determined on the basis of a single brood alone. However, this may result in unmanageable sample sizes, a problem that cannot be overlooked. Sampling more exposed males increases the chance of detecting a mutagenic event because the precision of the estimated mutation rate will be increased. Consequently it is necessary to maximise the numbers of treated males (recommended numbers are given in Section 9.4.1.1). However, this leads to a corresponding increase in the numbers of vials, since males must be kept separate from each other (Section 9.2.2).

9.4.1.1 *Important points concerning the protocol*

(a) The minimum number of males necessary to give the required number of treated sperm and thus exposed *X*-chromosomes must be selected carefully, taking into account losses due to mortality and human error which may occur during the experiment (see Section 9.4.1). At least 50 males are mated individually to 1–3 females for each concentration of test chemical per brood. The relative numbers of surviving flies in the control and treated samples which can be analysed determine the type of statistical test that can be applied to the data (see Section 9.4.4).

(b) Each individual vial must be labelled so that clusters can be traced back to individual treated males.

(c) A concurrent negative control group of male flies should be included and dosed with a vehicle control using the identical mode of administration to that used for the test compound. Once the procedures have become well established in a laboratory, it may be judged unnecessary to include a positive control. However, this working group recommends the use of positive controls from time to time to monitor the performance of the protocol.

(d) Doses of test agent should be selected carefully by using the results of toxicity screening tests (for methods see Mitchell & Combes, 1984). The top dose chosen must result in at least 50% survival and have minimal effects on fertility and the sex ratio. At least one other lower dose should be included in protocols for routine testing. To account for lethality, at least 100 males should be exposed per treatment.

(e) In order to ensure that the F_2 generation is comprised of progeny derived from a representative sample of original males, it is important to establish comparable numbers of sib (brother/sister) matings (Fig. 9.1) of F_1 flies from the offspring of each parental male. Failure to do this can exacerbate effects due to clusters of mutations. In order to counteract any clustering effects, Valencia *et al.* (1985) recommend that no more than 40 F_1 females be mated individually from each brood of each male. Furthermore, if more than two lethal mutations are observed from a single P_1 male, with a frequency greater than can be explained by chance alone, these authors assume this to be evidence for clusters and recommend that all the progeny from that male should be eliminated from the analysis (but see Section 9.2.3).

If low numbers of progeny are recovered in the F_2 generation vials, problems can arise in the interpretation of data. For example, if 12 flies in total are recovered in an F_2 vial, three wild type males would be expected due to equal ratios of the four classes of progeny. However, with such low numbers, an absence of males may be due to chance rather than to the fact that a recessive lethal mutation that occurred in the parental male was inherited. Bootman & Kilbey (1983) discuss this problem in more detail, and this working group agrees with their recommendations that any F_2 vial with less than 12 flies in total and lacking a wild type male should not be included in the scoring.

(f) No statistical analysis of the F_3 progeny is required as this generation is obtained simply to confirm the presence or absence of a recessive lethal in the corresponding F_2 vial.

9.4.2 Introduction to methods of analysis

The most appropriate method for analysis of data can be selected only when the data have been inspected. Usually the normal test (Section 9.4.3.4) will be appropriate for the analysis of results arising from Drosophila experiments where a large number of experimental units is involved (usually 1000—3000). However, the normal test will be inappropriate when numbers in the treated and control samples differ substantially and when the level of mutation in the control group is too small (Section 9.4.4).

The data will be in the form illustrated in Table 9.2 where M is the total number of mutants detected in both treatment groups and X_c and X_t respectively are the observed numbers of mutants in a total sample of N_c and N_t. Each unit has, independently of other units, an underlying probability of mutation p_c or p_t.

9.4.2.1 *Assumptions made and potential complications*

It should be remembered that there may not be a simple relationship between the dose of a compound administered and the concentration of the corresponding active chemical to which the target cells are exposed. This discrepancy may be due to unpalatable food or to the modulating effects of metabolism. In the case of unpalatability, the testing of lower doses of a compound may be advisable in order to maximise ingestion and the concomitant level of exposure of the target cells.

It is very difficult to make accurate comparisons between treatments of different chemicals. This is because each compound may be applied at a unique concentration which is dictated by its specific general toxicity and mutagenic potency. With single treatments of each compound, it is only possible to obtain information about relative mutagenic potencies when identical doses are used. However, such information is best obtained from the analysis of dose–response data based on a binomial distribution of probabilities. This topic has been reviewed by Finney *et al.* (1963) and Hewlett & Plackett (1979).

In assessing comparative mutation, either the difference $(p_t - p_c)$ or the ratio (p_t/p_c) is used. Three methods have been used for statistical analysis and three basic assumptions underlie each test. These are:

(a) the data fall into two mutually exclusive classes (mutagenic or non-mutagenic);

(b) each observation is independent of any other one; and

(c) each individual within a test or a control group has the same probability of giving rise to a positive observation.

Although assumption (a) is true, criterion (b) does not always hold due to the occurrence of clusters. Condition (c) may also be violated because each fly will have a unique level of tolerance to any dose of the test chemical. However, this level is undefinable and the resulting potential inaccuracy cannot be accounted for in any statistical model.

9.4.3 Examples of the recommended statistical tests

This section illustrates the computational procedures used in applying the Fisher's exact, the conditional binomial and the normal tests to data obtained from the sex-linked recessive lethal assay. In the descriptions of the tests that follow the data are presented in the general format given in Table 9.2.

9.4.3.1 *Fisher's exact test*

This test for a significant difference between the treatment and the control group is based on the relationship between the number of treatment mutations (X_t) and the margins (N_c, N_t, M and N–M) presented in Table 9.2. If X_c is unusually small, then the null hypothesis $H_0{:}P_t = P_c$ is likely to be rejected. In order to test this hypothesis, it is necessary to calculate the probability of actually observing X_c, given the marginal totals of the table from the relationship:

$$pr(X_t|N_c,N_t,M,N-M) = \frac{N_c!N_t!M!(N-M)!}{X_c!X_t!(N_c-X_c)!(N_t-X_t)!N!} \tag{3}$$

(where the left-hand side of the above equation (3) is mathematical notation for the probability of observing X_t mutations in the treatment group given the marginal totals)

After calculating this probability, the probabilities associated with the more extreme values of X_c (i.e. X_c-1, X_c-2, . . . , 0) given the same marginal totals have to be obtained in the same way. When all these probabilities are summed, the resultant probability (P) can be compared with the chosen critical probability (e.g. 0.05) for significance. An alternative to the above procedure for calculating P is provided by Finney *et al.* (1963) who give critical values of X_c at given N_c and N_t for significance at the 0.01 or 0.05 levels.

Consider the following example of mutation test data, based upon an illustration in Selby & Olsen (1981): (see Margolin *et al.*, 1983).

	Control	Treatment	Totals
Mutant X-chromosomes	2	4	6
Non-mutant X-chromosomes	30	15	45
Totals	32	19	51

For a one-sided test, the number of extreme tables is equal to the number of observed control mutants. Since the number (X_c) of control mutants in the above Table is 2, there will be two extreme tables, with values of control mutants of 1 and 0 (X_c-1, X_c-2, stopping at 0 as $X_c=2$). The other values are consequent on the marginal total numbers which remain fixed (i.e. $1+31=32$; $1+5=6$; $0+32=32$ and $0+6=6$). The set of more extreme tables (one-sided) for this example are presented below.

Extreme table for $X_c-1=1$

	Control	Treatment
Mutant X-chromosomes	1	5
Non-mutant X-chromosomes	31	14

Extreme table for $X_c-2=0$

	Control	Treatment
Mutant X-chromosomes	0	6
Non-mutant X-chromosomes	32	13

P can now be calculated by substituting into equation (3) the relevant numbers from the above three tables to give:

$$P=\frac{6!\ 45!\ 32!\ 19!}{51!}\left[\frac{1}{2!\ 4!\ 30!\ 15!}+\frac{1}{1!\ 5!\ 31!\ 14!}+\frac{1}{0!\ 6!\ 32!\ 13!}\right]$$
$$=0.1067+0.0207+0.0015=0.1289 \qquad (4)$$

where, for example, $6!=6\times5\times4\times3\times2\times1$ and $0!$ is defined to be unity. As $P=0.1289>0.05=\alpha$, the null hypothesis is not rejected and it is concluded that the data are consistent with the test compound being non-mutagenic.

9.4.3.2 *The conditional binomial test*

In this test, the null hypothesis H_0, conditional on there being X_t+X_c observed mutations, is that the ratio of control group mutations to mutations in the control and treatment groups together is binomial. The probability of observing X_c or more mutations from a total of X_c+X_t mutations, given that the relative proportion of mutations in this group under H_0 is p, is:

$$\sum_{r=k}^{X_c+X_t} {}^{(X_c+X_t)}C_r\, p^r(1-p)^{X_c+X_t-r} \qquad (5)$$

where, in the context of the conditional binomial test, p is a probability (Kastenbaum & Bowman, 1970).

The experimenter then will reject H_0 whenever $X_c > K$, where K is the critical-valued smallest integer such that:

$$\sum_{r=k}^{X_c+X_t} {}^{(X_c+X_t)}C_r\, p^r(1-p)^{X_c+X_t-r} \leqslant \alpha \qquad (6)$$

and α is the pre-assigned level of significance.

The heuristic equations (5) and (6) have been included to illustrate the theoretical basis of the test. They are not working equations and are therefore unnecessary for the actual application of the test. The reader who wishes to apply the test for analysis of data will find the working method described in Section 9.4.3.3 below.

9.4.3.3 *Example of the application of the conditional binomial test*

The following data are from Kastenbaum & Bowman (1970):

$X_t = 69$, $N_t = 262770$
$X_c = 15$, $N_c = 82980$

To test H_0 by use of the Kastenbaum–Bowman Tables (Kastenbaum & Bowman, 1970) an investigator will require to know the following:

(a) the total number of mutants, $n = (X_t + X_c)$, and
(b) an estimate of the probability of mutation in the treatment group assuming the null hypothesis (H_0) is true, i.e. $p_t = p_c$.

This estimate is denoted by p in the Kastenbaum–Bowman tables and is given by $N_t/(N_t + N_c)$.

Using the data from the above example, $n = (69 + 15) = 84$, and $p = 262770/(262770 + 82980) = 0.76$. For $\alpha = 0.05$, from the tables K is 71 (K is equivalent to c in the Kastenbaum–Bowman tables). This tabulated value represents the minimum number of sex-linked recessive lethal mutations in the treated group which permits rejection of the null hypothesis (H_0). Thus H_0 can be rejected if the value of X_t is greater than K and the compound can be considered mutagenic at the α value used. In this example, as X_t (= 69) is less than K(= 71), H_0 cannot be rejected at the 5% significance level and it is concluded that the data are consistent with the test compound being non-mutagenic.

9.4.3.4 *The normal test*

A binomial distribution with sample size n and probability of mutation p will have mean np and variance $np(1-p)$. Provided n is sufficiently large this distribution can be approximated reasonably well by a

Table 9.5. *The appropriateness of statistical tests*

Test	N^a	P_c
Fisher's exact	<100	$0.00 \leqslant P_c \leqslant 0.05$
Conditional binomial	<100	$0.05 \leqslant P_c \leqslant 0.95$
Conditional binomial	>100	$0.00 \leqslant P_c \leqslant 0.15$
Normal	>100	$N_c P_c \geqslant 2$

Note:
[a] $N = n_c = n_t$.

normal distribution with the same mean and variance. Mutagenicity can then be analysed by use of a normal test as described below.

The difference (D) between the treatment and control mutation relative frequencies,

$$D = (X_t/N_t) - (X_c/N_c) \tag{7}$$

is divided by an estimate of its standard error (SE) calculated under the assumption that the test agent is not a mutagen using the formula:

$$SE = \sqrt{[p\,(1-p)\,((1/N_c)+(1/N_t))]} \tag{8}$$

where $p = (X_t + X_c)/(N_t + N_c)$. The value of the ratio $Z = D/SE$ (see equation (1) Section 9.3.2.1) is then compared with a critical value z_α from standard normal probabilities to ascertain its significance level.

9.4.3.5 ***Example of the application of the normal test***

Given that $X_t = 11$, $N_t = 1720$

$X_c = 4$, $N_c = 1720$

we have $D = 11/1720 - 4/1720 = 0.00407$

$p = (11+4)/(1720+1720) = 15/3440 = 0.00436$

$SE = \sqrt{[0.00436(1-0.00436)((1/1720)+(1/1720))]}$

$= 0.00225.$

Therefore $Z = D/SE = 0.00407/0.00225 = 1.8089$.

At a 5% significance level, the critical value of z_α is 1.645, and, as $1.8089 > 1.645$, the null hypothesis is rejected. It is concluded that the experimental data are consistent with the test compound being mutagenic at the 5% significance level.

9.4.4 Comparison of statistical methods

Table 9.5 summarises experimental situations which are appropriate for the application of the three statistical tests which have been described. Where the sample (N) is less than 100, the normal test should not be used (Table 9.5). If the probability of control mutation is 0.05 or less, then the Fisher's exact test should be applied. However, should the probability exceed this figure, but be less than 0.95, then the conditional binomial test is the most appropriate method for analysis.

Margolin *et al.* (1983) have investigated experimental situations with sample sizes greater than 500 and a probability of mutation less than 0.05. This is the precise situation that is encountered when using Drosophila. Margolin *et al.* (1983) concluded that, for a ratio of sample sizes between 0.80 and 1.25 (which includes the recommended equality in sample sizes), the normal test is preferable to the conditional binomial. This assumes that, when the control sample size is multiplied by the probability of control mutation, the result is equal to or greater than 2 (i.e. $N_c p_c >= 2$). Thus, with a background mutation rate (p_c) of 0.0015, the required minimum sample size (N_c) is $2/0.0015 \doteq 1333$. If an investigator wishes to use the normal test it follows that it is necessary to analyse roughly comparable numbers of progeny in the control and treatment groups and to use sufficient control flies to satisfy the requirement above. The choice of test will therefore depend upon the experimental conditions.

Berkson (1978) and Upton (1982) have criticised the use of Fisher's exact test as a substitute for the normal test, since in practice the actual significance levels of tests (the effective levels) may be quite different from the nominal significance levels adopted *a priori.* This difference decreases the power of detection for positive effects (see Section 9.3.2.3). Although Fisher's exact test could be made more precise by use of a randomised version, this has little appeal in practice. Moreover, Suissa & Shuster (1985) point out that, for non-statisticians, the interpretation of results is more obvious when using the normal test, since it can be difficult to understand the underlying theory of Fisher's exact test, which is a conditional test.

Frei & Wurgler (1988) recently have published a slightly different approach to the analysis of data in which less emphasis is placed on allowing calculation of sample sizes needed to satisfy given values of the Type I and II errors (α and β).

9.4.5 Worked examples

Raw data obtained using the sex-linked recessive lethal mutation assay have been selected from the literature in order to illustrate some of the problems encountered when applying the recommended statistical tests. Table 9.6 presents results published by Parry & Sinclair (1985) for several compounds tested as part of the recent UKEMS collaborative trial. These authors used the protocol suggested by Mitchell and Combes (1984) and recommended in this chapter. The only difference in the protocols is the use

Table 9.6. *Raw data from sex-linked recessive lethal assay for four different compounds (pooled data)*[a]

Compound	μg/ml	N_c	N_t	(N_c+N_t)	(N_c/N_t)	(N_t/N_c)	P_c	X_c	X_t
BZD	1000	5381	3841	9222	1.40	0.71	0.0013	7	5
	2000	5381	6592	11973	0.82	1.23	0.0013	7	13
DAT	500	5381	3458	8839	1.56	0.64	0.0013	7	21
	1000	5381	6391	11772	0.84	1.19	0.0013	7	35
DAB	1000	5381	3946	9327	1.36	0.73	0.0013	7	32
	2000	5381	4303	9684	1.25	0.80	0.0013	7	77
CDA	200	5381	3899	9280	1.38	0.72	0.0013	7	23
	1000	5381	3567	8948	1.51	0.66	0.0013	7	30

Note:
[a] data from Parry & Sinclair (1985).
$P_c = X_c/N_c$.

Table 9.7. *Raw data from sex-linked recessive lethal assay for four different compounds (brood data)*[a]

Compound	μg/ ml	brood	N_c	N_t	(N_c+N_t)	(N_c/N_t)	(N_t/N_c)	P_c	X_c	X_t
BZD	1000	1	1398	1201	2599	1.16	0.86	0.00072	1	2
		2	1267	867	2134	1.46	0.68	0.00079	1	1
		3	1401	931	2332	1.50	0.66	0.0014	2	2
		4	1315	842	2157	1.56	0.64	0.0023	3	0
DAT	1000	1	1398	1538	2936	0.91	1.10	0.00072	1	5
		2	1267	1796	3063	0.71	1.42	0.00079	1	11
		3	1401	1431	2832	0.98	1.02	0.0014	2	13
		4	1315	1626	2941	0.81	1.24	0.0023	3	9
DAB	1000	1	1398	931	2329	1.50	0.67	0.00072	1	7
		2	1267	876	2143	1.45	0.69	0.00079	1	11
		3	1401	1013	2414	1.38	0.72	0.0014	2	8
		4	1315	1126	2441	1.17	0.86	0.0023	3	6
CDA	200	1	1398	1025	2423	1.36	0.73	0.00072	1	1
		2	1267	945	2212	1.34	0.75	0.00079	1	8
		3	1401	887	2288	1.58	0.63	0.0014	2	7
		4	1315	1042	2357	1.26	0.79	0.0023	3	7

Note:
[a] data from Parry & Sinclair (1985)
$P_c = X_c/N_c$.

of four broods, instead of three (Table 9.7). In order to select the most appropriate statistical test, the total sample size $(N_c + N_t)$ is computed. As this population size exceeds 100 in all cases, Fisher's exact test is inappropriate (Table 9.5). Although there are seven mutants in the pooled

Table 9.8. *Statistical analysis of pooled data using the conditional binomial test*[a]

Compound	μg/ml	P^b	N^c	X_t	value of K and significance at: $P<0.05$	$P<0.01^d$
BZD	1000	0.42	12	5	9 (*NS*)	10 (*NS*)
	2000	0.55	20	13	16 (*NS*)	17 (*NS*)
DAT	500	0.39	28	21	16 (*S*)	18 (*S*)
	1000	0.54	42	35	29 (*S*)	31 (*S*)
DAB	1000	0.42	39	32	22 (*S*)	25 (*S*)
	2000	0.44	84	77	45 (*S*)	49 (*S*)
CDA	200	0.42	30	23	18 (*S*)	20 (S)
	1000	0.40	37	30	21 (*S*)	23 (*S*)

Notes:
[a] obtained from Table 9.6.
[b] $P=(N_t/N_t+N_c)$.
[c] $N=(X_t+X_c)$.
[d] significance from Kastenbaum–Bowman tables with significance where $X_t>K$ at different values of α indicated in parentheses.

control population (Table 9.6), the ratios of the control and treated populations fall within the range 0.80–1.25 only in the cases of benzidine (BZD) at 2000 μg/ml, diaminoterphenyl (DAT) at 1000 μg/ml and dimethylaminoazobenzene (DAB) at 2000 μg/ml. Thus, although these data can be analysed by the normal test, the other results should be subjected to the conditional binomial test. The outcome of this analysis is presented in Tables 9.8, 9.9. The conditional binomial test is appropriate (Table 9.5) since the probability of spontaneous mutation (p_c) is <0.15 in every test.

For the differences between treatment and controls to attain significance

Table 9.9. *Statistical analysis of pooled data for three compounds using the normal test*[a]

Compound	μg/ml	D	P	SE	$Z(D/SE)$	significance[b] ($p<0.05$)
BZD	2000	0.00067	0.0017	0.00075	0.89	(*NS*)
DAT	1000	0.0042	0.0036	0.0011	3.79	(*S*)
DAB	2000	0.017	0.0087	0.0019	8.75	(*S*)

Notes:
[a] data from Table 9.6 *see* Sections 9.4.3.4 and 9.4.3.5 for methods of working out values.
[b] significant where $z>$ critical value $z_\alpha=1.645$.

Table 9.10. *Statistical analysis of brood data for three different compounds in the sex-linked recessive lethal assay using the conditional binomial test*[a]

Compound	μg/ml	Brood	P^b	N^c	X_t	value of $K(c)$ and significance at $P<0.05$	$P<0.01^d$
BZD	1000	1	0.46	3	2	*nv*	*nv*
		2	0.41	2	1	*nv*	*nv*
		3	0.40	4	2	4 (*NS*)	*nv*
		4	0.39	3	0	*nv*	*nv*
DAT	1000	1	0.52	6	5	6 (*NS*)	*nv*
		2	0.59	12	11	11 (*BS*)	12 (*NS*)
		3	0.51	15	13	12 (*S*)	13 (*BS*)
		4	0.55	12	9	10 (*NS*)	11 (*NS*)
DAB	1000	1	0.40	8	7	6 (*S*)	7 (*BS*)
		2	0.41	12	11	9 (*S*)	10 (*S*)
		3	0.42	10	8	8 (*BS*)	9 (*NS*)
		4	0.46	9	6	8 (*NS*)	9 (*NS*)
CDA	200	1	0.42	2	1	*nv*	*nv*
		2	0.43	9	8	7 (*S*)	8 (*BS*)
		3	0.39	9	7	7 (*BS*)	8 (*NS*)
		4	0.44	10	7	8 (*NS*)	9 (*NS*)

Notes:
[a] raw data from Table 9.7.
[b] $P=N_t/(N_t+N_c)$.
[c] $N=(X_t+X_c)$
[d] (*S*), significant at *P* level indicated- (*NS*), not significant; (*BS*), borderline significance; *nv*, no value given in Kastenbaum–Bowman tables.

using this test (Section 9.4.3.2), the counts for X_t must be greater or equal to the respective tabulated values of *K* (Kastenbaum & Bowman, 1970). Comparing the values in Tables 9.6 and 9.8 confirms that all tests were significant at both values of α (0.05 and 0.01) except when BZD was tested at either concentration. These conclusions are endorsed by the analyses undertaken with the normal test where z exceeds the critical value z_α of 1.645 in the case of DAT at 1000 μg/ml and DAB (2000 μg/ml) but not for BZD.

The individual results for brooding, obtained with the same compounds, are included in Table 9.7 and the results for analysis with the conditional binomial test are presented in Table 9.10. When the tabulated values of *K* are compared with the respective figures for mutants in the treated population (X_t), it is clear that no significant mutagenicity was induced in any brood by BZD. In fact, in all but one test, the calculated figures for p (N_t/N_t+N_c) and (X_c+X_t) are too low for values of *K* to be provided in the

Kastenbaum–Bowman tables at either of the significance levels ($\alpha = 0.05$ and 0.01) that are usually employed. Consequently, there is no value which X_t must exceed to be significant, so strictly speaking the significance of the results cannot be assessed at such values of α. Under these circumstances most investigators would consider such results to be insignificant, but in reality, the experimental design is insufficiently rigorous to provide data that can be subjected to statistical analysis. Therefore, the mutagenicity of such a compound cannot be assessed without further experimentation using an improved protocol. At $\alpha = 0.1$ the number of mutants (X_t) is less than the corresponding value of K so, at this α, the data are insignificant.

The mutagenicity of DAT was just significant in brood 2 (Tables 9.7 and 9.10), where the values of X_t and K at $\alpha = 0.05$ were identical, whilst in brood 3 the numbers of mutants in the treated population are equal to the critical value K at $\alpha = 0.01$ and hence DAT is considered mutagenic. However, when tested in brood 4, the number of mutants in the treatment group was less than K and therefore based on the evidence using this brood DAT would be classed non-mutagenic. By the same criteria, DAB was classed as mutagenic in brood 1 and brood 2 ($\alpha = 0.01$) and in brood 3 ($\alpha = 0.05$) but would be classed as non-mutagenic on the evidence from brood 4. The results of cyanodimethylaniline (CDA) were significant and just significant respectively for broods 2 and 3 only, with X_t the same as K at $\alpha = 0.05$, for brood 3. At $\alpha = 0.01$ they were just significant (brood 2) and non-significant (brood 3).

The following data meet the criteria for analysis by the normal test: DAT (broods 3 and 4) and DAB (brood 4). CDA (brood 4) is included since the ratios of n_c and n_t were only just outside the recommended range and because the value of X_t was 1 less than the tabulated minimal value of K at $\alpha = 0.05$. The results of the test (Table 9.11) using $\alpha = 0.05$ agree with those obtained using the conditional binomial test (Table 9.10). With CDA (brood 4), the value of z is 1.63 which is less than the critical value of z_α (1.645). This indicates that the test result is not significant at the stated level of α.

Table 9.12 includes routine, unpublished data observed when testing the positive control ethylmethane sulphonate (EMS) and three coded compounds, X, Y and Z. These results represent pooled counts from three separate broods. All of the tests, except EMS, are amenable to analysis using the normal test, the results of which are presented in Table 9.13. The values for z are < 1.645 in all tests. Thus the three compounds can be classed as non-mutagenic even when pooled data were analysed. When the EMS data are tested using Kastenbaum–Bowman tables, the respective values for n, p and K were 35, 0.26 and 16, respectively. As the value for X_t is more

Table 9.11. *Statistical analysis of brood data for DAT, DAB and CDA using the normal test*

Compound	μg/ml	brood	D	P	SE	$Z(D/SE)$	significance ($P<0.05$)[b]
DAT	1000	3	0.0077	0.0053	0.0027	2.85	(*S*)
		4	0.0033	0.0041	0.0024	1.38	(*NS*)
DAB	1000	4	0.0030	0.0037	0.0025	1.20	(*NS*)
CDA	200	4	0.0044	0.0042	0.0027	1.63	(*NS*)

Notes:
[a] data from Table 9.7.
[b] significant where $z>$ critical value $z_{\alpha}=1.645$.

Table 9.12. *Raw data from sex-linked recessive lethal assay for three different compounds and ethylmethane sulphonate (pooled data)*[a]

Comp.	conc.	N_c	N_t	(N_c+N_t)	(N_c/N_t)	(N_t/N_c)	P_c	X_c	X_t
X	low	1850	1755	3605	1.05	0.95	0.0011	2	4
Y	low	2616	2628	5244	1.00	1.00	0.00076	2	4
	high	2616	2582	5198	1.01	0.99	0.00076	2	4
Z	low	2332	2293	4625	1.02	0.98	0.0017	4	5
	high	2332	2314	4646	1.01	0.99	0.0017	4	4
EMS	0.005%	1850	666	2516	2.78	0.36	0.0011	2	33

Notes:
[a] data from Parry & Sinclair (1985).
$P_c=X_c/N_c$

Table 9.13. *Statistical analysis of pooled data for three coded compounds using the normal test*[a]

Compound	μg/ml	D	P	SE	$Z(D/SE)$	significance ($P<0.05$)[b]
X	low	0.0012	0.0017	0.0014	0.88	(*NS*)
Y	low	0.00076	0.0011	0.00093	0.81	(*NS*)
	high	0.00078	0.0012	0.00094	0.83	(*NS*)
Z	low	0.00047	0.0019	0.0013	0.36	(*NS*)
	high	0.000013	0.0017	0.0012	0.011	(*NS*)

Notes:
[a] data from Table 9.12.
[b] significant where $z>$ critical value $z_{\alpha}=1.645$.

Table 9.14. *Sex-linked recessive lethal data for compound Y at low dose (brood 1) tested in the conditional binomial and normal tests*

									K(c)						
N_c	X_c	N_t	X_t	(N_c+N_t)	(N_c/N_t)	(N_t/N_c)	N	(N_t/N_c+N_t)	$P<0.05$	$P<0.10$	D	P	SE	Z	sign ($P<0.05$)
866	0	863	4	1729	1.00	1.00	4	0.50	*nv*	4	0.0046	0.0023	0.0023	2.00	(*S*)

than double the value for K the results for EMS induction in this test are highly significant.

Spurious results can be obtained through inappropriate use of statistical tests. Table 9.14 shows data obtained for compound Y when tested on brood 1. Since the ratio of N_c and N_t is 1 it might be considered correct to use the normal test in which case a value for z of 2.00 is obtained. This observed value of z is > 1.645 (the critical value) and therefore it would be concluded that the result of the test was positive. However, this analysis should not be applied since $p_c N_c = < 2$. If the conditional binomial test is used, no tabulated number for K is obtained for the values of N and P at $\alpha < 0.05$. If α is set at $p < 0.1$ then K is 4 which, in this case, is the same as X_t. Thus the data for compound Y actually are on the borderline of significance at the highest normally accepted critical probability value.

Whenever possible, analysis of data obtained in an assay should be based on the same statistical test. However, as can be seen from these worked examples, this is not always feasible. Experiments should be designed bearing in mind the recommendations laid out in this chapter and with regard to the capacity of the laboratory to cope with the required protocol.

9.5 CONCLUSIONS

(a) When the sex-linked recessive lethal assay is undertaken, the protocol mentioned in previous sections should be considered. It is important to maintain the numbers of exposed males at levels which will maximise the resolving power of the statistical assay. At the completion of the experiment, the most appropriate statistical test should be selected on the basis of the total sample size and the control mutation frequency which must be obtained from experiments conducted concurrently.

(b) A test based on the normal approximation to the binomial distribution provides the most sensitive means of distinguishing between mutagenic and control treatments. This assumes that the necessary criteria concerning sample sizes and mutation rates are met.

(c) Although other tests for handling 2×2 tables have been recommended (Upton, 1982) as improvements on the normal test under circumstances where the conditions in Section 9.4.4 are fulfilled, only small numbers of progeny can be scored and none of these tests offers any more than a marginal improvement over the normal test. Therefore, the working group felt unable to recommend any one of them in preference to the normal test.

(d) Where sample sizes and ratios permit, and provided that the expected number of mutations ($N_c p_c$) is at least two in the control population, the normal test is the recommended form of statistical analysis to apply. Where these criteria are unfulfilled, then analysis should be undertaken using the conditional binomial test. The Fisher's exact test is not recommended for the analysis of Drosophila mutation data because the number of experimental units is too large.

9.6 REFERENCES

Berchtold, W. (1975). Comparison of the Kastenbaum–Bowman and Fisher's exact test, *Arch. Genetik.*, **48**, 151–7.

Berkson, J. (1978). In dispraise of the Exact test, *Journal of Statistical Planning and Inference*, **2**, 27–42.

Bootman, J. & Kilbey, B.J. (1983). Recessive lethal mutations in Drosophila, in: *UKEMS Sub-committee on Guidelines for Mutagenicity Testing*. Report, Part I, Basic Test Battery, Ed. B.J. Dean, United Kingdom Environmental Mutagen Society, Swansea, pp. 103–17.

Crow, J.F. & Temin, R.G. (1964). Evidence for the partial dominance of recessive lethal genes in natural populations of Drosophila, *American Naturalist*, **98**, 21–33.

Engels, W.R. (1979). The estimation of mutation rates when premeiotic events are involved, *Environmental Mutagenesis*, **1**, 37–43.

Feller, W. (1971). *An Introduction to Probability Theory and its Applications*, Wiley, New York.

Finney, D., Latscha, R., Bennett, B., Hou, P. & Pearson, E. (1963). *Tables for Significance Testing in a 2 × 2 Contingency Table*, Cambridge University Press, Cambridge.

Frei, M. & Wurgler, F.E. (1988). Statistical methods to decide whether mutagenicity test data from Drosophila assays indicate a positive, negative or inconclusive result. *Mutation Research*, **203**, 297–308.

Hewlett, P.S. & Plackett, R.L. (1979) *The Interpretation of Quantal Responses in Biology*, Arnold, London

Kastenbaum, M.A. & Bowman, K.O. (1970). Tables for determining the statistical significance of mutation frequencies, *Mutation Research*, **9**, 527–49.

Lee, W.R., Abrahamson, S., Valencia, R., Von Halle, E.S., Wurgler, F.E. & Zimmering, S. (1983). The sex-linked recessive lethal test for mutagenesis in *Drosophila melanogaster* – A report of the US Environmental Protection Agency Gene-Tox Program, *Mutation Research*, **123**, 183–279.

Lehman, E.L. (1959). *Testing Statistical Hypotheses*, Wiley, New York.

Margolin, B.H., Collins, B.J. & Mason, J.M. (1983). Statistical analysis and sample size determinations for mutagenicity experiments with binomial responses, *Environmental Mutagenesis*, **5**, 705–16.

Mitchell, I.deG. & Combes, R.D. (1984). Mutation tests with the fruit fly *Drosophila melanogaster*, In: *Mutagenicity Testing – A Practical Approach*, ed. S. Venitt & J.M. Parry, IRL, Oxford, pp. 149–85.

Muller, H.J. (1952). The standard error of the frequency of mutants some of which are of common origin, *Genetics*, **37**, 608.

Parry, J.M. & Sinclair, P.B. (1985). An evaluation of the genotoxicity of BZD,

DAT, DAB and CDA in the sex-linked recessive lethal test using *Drosophila melanogaster*, In: *Comparative Genetic Toxicology*, ed. J.M. Parry & C.F. Arlett, Macmillan, London, pp. 499–505.
Selby, P.B. & Olson, W.H. (1981). Methods and criteria for deciding whether specific-locus mutation rate data in mice indicate a positive, negative or inconclusive result, *Mutation Research*, **83**, 403–18.
Sokal, R.R. & Rohlf, F.J. (1981). *Biometry*, 2nd Edn, Freeman, San Francisco.
Suissa, S. & Shuster, J. (1985). Exact conditional sample sizes for the 2 × 2 binomial trial, *Journal of the Royal Statistical Society A*, **148**, 317–27.
Upton, G.J.G. (1982). A comparison of alternative tests for the 2 × 2 comparative trial, *Journal of the Royal Statistical Society A*, **145**, 86–105.
Valencia, R., Mason, J., Woodruff, R. & Zimmering, S. (1985). Chemical mutagenesis testing in Drosophila. III. Results of 48 coded compounds tested for the National Toxicology Program, *Environmental Mutagenesis*, **7**, 325–48.
Vogel, E. (1985). The Drosophila somatic recombination and mutation assay using the white-coral somatic eye colour system, in: *Progress in Mutation Research* Vol 5, Elsevier, Amsterdam, pp. 313–17.
Wurgler, F.E., Sobels, F.H. & Vogel, E. (1977). Drosophila as assay system for detecting genetic changes, In: *Handbook of Mutagenicity Test Procedures*, ed. B.J. Kilbey, M. Legator, W. Nicols & C. Ramel, Elsevier, Amsterdam, pp. 335–73.
Wurgler, F.E. Sobels, F.H. & Vogel, E. (1984). Drosophila as an assay system for detecting genetic changes, In: *Handbook of Mutagenicity Test Procedures*, 2nd Edn., ed. B.J. Kilbey, M. Legator, W. Nichols & C. Ramel, Elsevier, Amsterdam, pp. 555–613.

APPENDIX 1
Glossary of terms

Algorithm A sequence of logical and mathematical steps by which problems can be solved.

Aligned rank order test A non-parametric statistical test which correlates aligned concentration means with predicted concentration rankings.

Analysis of variance (ANOVA) A statistical test that investigates differences between the mean of several samples.

Arithmetic mean A central value for a variable, often referred to as the average.

Binomial distribution The distribution defined by a sequence of probabilities, which are the successive terms of the binomial expansion of $(p+q)^n$.

Bonferroni corrections Corrections used to adjust multiple comparisons to reduce the risk of Type I errors. Nominal probability α is divided by the number of possible comparisons (n) and value of α/n is taken as critical value.

Chi-square (χ^2) distribution The distribution of the sum of squares of independent normal deviates.

Coefficient of variation The ratio of the population standard deviation to the population mean.

Confidence limits Measures of the precision with which estimates of a population from samples are determined.

Critical value The value of a test statistic beyond which (above or below depending on the test) any hypothesis is rejected.

Degrees of freedom The number of independent comparisons that can be made among a series of observations.

***F*-test** A ratio of variance between classes and within classes that test whether all classes derive from the same population.

Hypergeometric probability Hypergeometric distribution is the distribution equivalent to the binomial distribution except that sampling is from a finite population without replacement.

Iteratively reweighted least squares A method of least squares for finding estimates of parameters where the information or weight given to the individual data is repeatedly adjusted based upon previous analysis.

Likelihood function The function relating an unknown parameter to observations from which it can be estimated. The inverse of a probability function.

Likelihood ratios Ratio of two likelihoods used as a statistic to measure the degree of agreement between sampled and expected frequencies.

Linearisation of dose–response curve The transformation of the dose, or the response, or both, so that the results of a dose–response study can be plotted in a graphical form as an approximate straight-line relationship.

Linear trend A linear relationship between the response and the administered dose or treatment.

Log-likelihood ratios Statistic calculated from likelihood ratio statistic by multiplying the logarithm of the likelihood ratio by twice the natural logarithm of 10 (i.e. ln 10). This statistic is numerically similar to X^2.

Maximum likelihood A principle by which estimates of unknown parameters are chosen so as to maximise the likelihood of the observed data.

Monte Carlo method A method that uses random numbers to represent uncertain events in simulations.

Monotonically increasing dose–response A dose–response in which the response at each high dose is as great or greater than that at the preceding dose level.

m-statistic The ratio of the sample variance to the mean, or the ratio of the chi-squared statistic to the degrees of freedom.

n Number of observations or items in a sample.

Normal distribution The most common of the theoretical distributions which gives a bell-shaped probability curve.

Normalisation of Data The transformation of data from a non-normal distribution to one where the data fits a normal distribution.

Null hypothesis The hypothesis that there is no difference between the two or more populations/conditions that are under examination.

Orthogonal Comparisons between observations, whose values are uninfluenced by changes in each other. Also called independent comparisons.

p The probability of success in a binomial population.

Poisson distribution A convenient approximation to the binomial distribution when n is very large and p tends to zero.

Population standard deviation (σ) A measure of the variation within a population.

Population variance (σ^2) The square of the population standard deviation (σ).

q The probability that an event will not occur in a binomial population.

Ranking Arranging all the values in a data set from lowest to highest (or from highest to lowest). The arranged data is then assigned the rank (1, 2, 3) designating the order in which they occur.

Regression The relationship of the dependent variable (Y) to the independent variable (X).

Regression coefficient The coefficient representing the rate of change of the dependent variate on the independent variate.

Sample standard deviation (s) An estimate of the measure of variation within a population.

$$s = \sum \frac{(x - \bar{x})^2}{n - 1}$$

Sample variance (s^2) The square of the sample standard deviation (s)

Stabilisation of variance A transformation designed to make the variance almost independent of the mean.

Test statistic A figure obtained from condensing data in a way that represents the population from which the observations were drawn.

Transformation Where basic distributional assumptions fail, the conversion of observations from one scale to another to permit standard analysis.

Weighted linear regression Regression analysis based on minimising a weighted sum of squares, derived by weighting each observation with the inverse of the associated variance (particularly useful where there is a good fit at low values but a poor fit – greater variance – at high values).

APPENDIX 2
Statistical software

M.R. THOMAS

The 1980s have seen sustained growth in the number of statistical packages, especially those available on microcomputers. Neffendorf (1983) provides a list of packages for micros, and the SGCSA (Study Group on Computers in Survey Analysis) maintain a register, price £4, of statistical software. These software lists are compiled from information received from package authors and are not intended to provide critical reviews of the products. It is unfortunate that some of this software is of very poor quality. The author has recently reviewed a package, distributed by a reputable scientific publisher, which boasted an impressive list of facilities for statistical analysis. Investigation revealed that the program could not calculate a mean or standard deviation correctly, all results were printed out to two decimal places (not much use for mutation frequencies), transformation of variables was almost impossible and program crashes resulted in lost data in over 50% of sessions. The package was deservedly cheap, but its purchase would be costly in terms of incorrect results, lost data, wasted effort and personal frustration.

One of the most useful developments in microstatistics software has been the arrival of micro versions of the well-known mainframe packages. With the exception of GENSTAT, all of the major mainframe packages are now available on PC compatible machines. Purchasers of microsoftware who lack detailed technical and product knowledge would be well advised to avoid the very cheap packages, and focus their attention on these mainframe migrants, and a very few other packages with proven track records, for example STATGRAPHICS, or INSTAT.

The choice of reliable statistical software for the older CPM micros is far more restricted. MINITAB is the only mainframe package which I know to have made the transition to 8-bit machines running CPM80 or CPM86. MASS was developed for CPM machines, and has received favourable reviews (Huson, 1985). NANOSTAT is available on CPM machines. The author has no personal experience of it and has not seen any independent reviews, but it is marketed by an internationally known centre of excellence in statistics: the Medical Statistics Department at the London School of Hygiene & Tropical Medicine. INSTAT is available for the BBC micro.

Further discussion will be restricted to software available on mainframes and on PC compatible machines; it will centre on the major packages: GENSTAT, SAS, BMDP, SPSS, MINITAB and STATGRAPHICS. No package in this list is uniquely better than the others; lengthy and futile discussion of their relative merits is one of the minor vices of statisticians. They may be differentiated in at least three directions: ability to perform complex data manipulation, range of statistical techniques accommodated, and ease of use. The best package for an individual scientist will depend on his needs in all of these directions, and upon other local factors. For example, scientists working in an organisation with a statistical consultancy will receive far better support if they use the same package as the statisticians.

General comments will be made about each of these packages, and they will be compared in terms of the provision which they make for some of the techniques advocated in this volume: weighted regression and weighted analysis of variance, multiple comparison procedures, trend tests and orthogonal polynomial contrasts, generalised linear models, Mann–Whitney *U*-tests and Fisher's exact test. Independent reviews of some of these packages have been published as follows: SAS, Beutel (1983), Thomas (1983); GENSTAT, Huson (1982); SPSS (PC version), Tucker (1985); and MINITAB, Horgan (1986), Shoesmith (1984).

GENSTAT is perhaps the most flexible of the statistical packages in terms of its ability to handle sophisticated statistical problems. It is a full programming language which enables expert users to implement virtually any statistical technique. Complicated data management is more difficult with GENSTAT than with SAS, SPSS and perhaps BMDP and PSTAT. For non-statisticians the main drawback has been the uncompromising style of the manual, which many would-be users find incomprehensible. Alvey *et al.* (1982) provide a simpler introduction to the package, but learning GENSTAT still requires a considerable effort. GENSTAT makes excellent provision for weighted and unweighted analysis of variance and regression, and for the estimation of orthogonal polynomials. It has poor provision for non-parametric tests, and will not easily perform Fisher's exact test. GENSTAT is the only one of these packages to make direct provision for generalised linear models.

SAS has a slightly more restricted range of packaged techniques than GENSTAT or BMDP, but, like GENSTAT, it provides tools to enable the expert user to perform virtually any statistical analysis. SAS is excellent at data management, so much so that statisticians have now become minority users of SAS. The product finds most of its market in commercial DP applications, where its statistical capabilities are ignored. It has excellent graphics. SAS is easier for the novice than GENSTAT, but the manuals are not an ideal introduction. The SAS institute provide a range of excellent training courses, but these are geared to private sector budgets. SAS will cope with all of the analyses, but needs expert coaxing to estimate generalised linear models. Although orthogonal polynomials may be estimated, the user must specify the appropriate coefficients; non-statisticians will find this difficult.

BMDP has a distinguished record in statistical innovation; especially in the area of multivariate analysis. It provides a wide range of specialised procedures, rather than a flexible tool for developing new techniques.

Although it is perfectly capable of performing the transformations recommended in this volume, its data management facilities are considerably less than those of SAS. The package will allow weighted and unweighted regression, and analysis of variance. In some of its analysis of variance procedures it is possible to estimate orthogonal polynomials, but, as with SAS, it is necessary for the user to enter the appropriate coefficients. BMDP makes good provision for non-parametric analysis.

MINITAB is a very popular interactive package. It provides good facilities for transforming data, but it is not intended to manage complicated data structures. The package is easy to use for simple analyses; it may be used to implement more complicated applications, but it then demands considerable statistical sophistication. The package provides weighted regression, but not (directly) weighted analysis of variance. It makes no direct provision for multiple comparison procedures.

SPSS has undergone considerable improvement, and SPSS-X and SPSS-PC now provide much better data management than the old SPSS. It is unlikely to be the choice of many statisticians, but it provides a comprehensive range of techniques and good data manipulation. Non-statisticains find it easy to master the package. All of the techniques mentioned above, with the exception of generalised linear models, are well catered for in SPSS.

Some authors in this volume have mentioned GLIM. GLIM was developed as a tool for statisticians, and people without statistical support should probably avoid it. GLIM is based on the concept of a generalised linear model, which in the author's opinion is the most significant advance in applied statistics of the past two decades. It provides a common theoretical and computational framework for previously distinct areas of applied statistics, such as linear models, quantal response models and log-linear models for contingency tables. GLIM provides the expert user with a concise and powerful data analysis tool.

It has been seen that the major differences between packages are in their facilities for complicated data management and for sophisticated statistical analysis. Most of the methods of analysis described here are relatively simple, and the test systems concerned do not generate large and complicated data structures. Most of the well-known statistical packages should prove adequate; the main danger lies in the cheaper packages written especially for micros.

CONTACT ADDRESSES FOR SOFTWARE

Register of Statistical Software, David Cable, SGCSA Treasures, c/o Central Statistical Office, Government Offices, Great George Street, London SW1P 3AQ

BMDP: Statistical Software Ltd, Cork Farm Centre, Dennehy's Cross, Cork

GENSTAT: The Statistical Package Co-ordinator, The Numerical Algorithms Group Ltd, NAG Central Office, Mayfield House, 256 Banbury Road, Oxford OX2 7DE

INSTAT: Statistical Services Centre, Department of Applied Statistics, University of Reading, Reading RG6 2AN

MASS: Mass Co-ordinator, Department of Statistics, Birkbeck College, Malet Street, London SW1

SAS: SAS Software Ltd, Wittington House, Henley Road, Medmenham, Marlow, Buckinghamshire SL7 2EB

SPSS: SPSS UK Ltd, London House, 243–53 Lower Mortlake Road, Richmond, Middlesex TW9 2LL

STATGRAPHICS: Mercia Software Ltd, Aston Science Park, Love Lane, Birmingham

REFERENCES

Alvey, N., Galwey, N. & Lane, P. (1982) *An Introduction to GENSTAT* Academic Press, London.

Beutel, P. (1983). SAS as a central statistical system. *The Professional Statistician*, **2**, 22–4.

Horgan, J. (1986). New developments in MINITAB. *The Professional Statistician*, **5**, 12–13.

Huson, L. (1982). GENSTAT – A user's view. *The Professional Statistician*, **1**, 10.

Huson, L. (1985). MASS (microprocessor applied statistics system). *The Professional Statistician*, **4**, 50–2.

Neffendorf, H. (1983). Statistical Packages for Microprocessors. *The Professional Statistician*, **2**, 8–11.
Shoesmith, E. (1984). MINITAB and SNAP reactions. *The Professional Statistician*, **3**, 10–11.
Thomas, M.R. (1983). A user review of SAS. *The Professional Statistician*, **2**, 26–7.
Tucker, C.E. (1985). SPSS PC – A user's view. *The Professional Statistician*, **4**, 15–16.

INDEX